Unemployment: Choices for Europe

Monitoring European Integration 5

BACKSTAGE AT
The Dean Martin Show

LEE HALE
with
Richard D. Neely

Thorndike Press • Thorndike, Maine

p. v: Dean at home base
p. vi: Lee Hale (*far left, smiling*) with Gail Martin, Dean, Tina Sinatra, Dino Martin, Nancy Sinatra, Frank Sinatra, and Frank Sinatra Jr. at orchestra rehearsal
p. ix: Frank Sinatra, Jeanne Martin, and Dean. Frank and Dean sport identifiers spoofing the ones Jack Halloran and I wore when we stood in as the two of them during rehearsals.
p. 18: Dean sliding down his fireman's pole on the set of *The Dean Martin Show*

All photographs by permission of Greg Garrison Productions

Published in 2001 by arrangement with DHS Literary Agency.

Thorndike Press Large Print Biography Series.

The tree indicium is a trademark of Thorndike Press.

The text of this Large Print edition is unabridged.
Other aspects of the book may vary from the original edition.

Set in 16 pt. Plantin by Minnie B. Raven.

Printed in the United States on permanent paper.

Library of Congress Cataloging-in-Publication Data

Hale, Lee.
    Backstage at the Dean Martin show / Lee Hale, with Richard D. Neely.
       p. cm.
    Originally published: Dallas : Taylor Pub., 2000.
    ISBN 0-7862-3234-X (lg. print : hc : alk. paper)
    1. Dean Martin show.   2. Large type books.   I. Neely, Richard D.   II. Title.
PN1992.77.D43 H35   2001
791.45'72—dc21   00-054467

Dedicated to the guest stars, singers and dancers, writers and choreographers, musicians and arrangers, cameramen and sound crew, stage managers and stagehands, costume and set designers, technicians and editors, secretaries and the cue card boys on our show, all of whom loved Dean.

"I've known Lee Hale for some time now (he co-produces our 'Ella' award shows for the Society of Singers) and through the years he's had countless wonderful stories to tell about the incredibly entertaining *The Dean Martin Show*. Now that he's put those stories into a book, I love it — I know I share that with all my friends who know Lee and knew Dean and loved his show so much. My husband, Henry Mancini, was a great fan, too."

Ginny Mancini

Foreword

Dear Reader,

How very pleased and grateful I am to Lee Hale for taking us all on such a joyful stroll down memory lane with our incomparable Dean Martin.

Dean would have been awed by such a task. As I am. What wonderful memories, Lee! And your telling of them is done with such humor, insight, and caring.

As far as I'm concerned, the first four years of *The Dean Martin Show* were the very best body of work that Dean ever did. Or anyone else, for that matter. That the show never did win an Emmy escapes me to this day! Did he make it appear too easy? Did they think he simply winged it? What fools!

Dean worked so terribly hard. The monumental pressure of the success or failure of each show eventually rested clearly on his shoulders and his alone. And he knew that. The fact that so few of his peers seemed to realize that is simply one more indication of his brilliance on that TV stage!

I suppose we at home took it in stride. After all, wasn't he always fun *and* funny in our own everyday living? And it wasn't easy to be a star at 601 Mountain Drive — everybody in the family thought they were! There wasn't a straight man to be found. Dean loved that.

It was so enjoyable for me to read Lee's account of all the inside happenings, both big and small, because I had never gone to a taping. I knew it would break his concentration. And what business did I have there, anyway?

All of our children and grandchildren (of which there are six and ten, respectively) will delight in this journal — a peek into Dean's world, his enchanted world. They all adored him and he gave them the happiest of childhoods.

As for me — well, I never considered anyone to be as talented as my husband. I loved him with all my heart. He was a true gentleman in every sense of that word.

There will never be another Dean Martin. I'm certain his multitude of talents will remain in memories for many years.

And aren't we all very fortunate that he passed through our lives? And left us smiling? I know I am.

Thank you for these memories, Lee.

With love,
Jeanne Martin

Contents

Introduction

When variety shows were the staple of TV viewing during the "Golden Years of Television," their patterns were amazingly similar. There was a well-known performer as host, usually a singing star, surrounded by a company of singers, dancers, and comics and featuring big-name stars as weekly guests.

Each show had a producer, director, choreographer, a staff of writers, and a musical conductor. They also had the guy who put all the musical elements together. On *The Perry Como Show*, that fellow was Ray Charles (the *other* Ray Charles). Danny Kaye and Dinah Shore had Earl Brown. Anyone who watched *Laugh-In* remembers the considerable musical contributions of Billy Barnes. Stan Freeman, Artie Malvin, and Ken and Mitzi Welch did it for Carol Burnett. What they did was come up with musical ideas, write music and lyrics, choose songs for star and guests, work out routines, and rehearse the whole company. It amounted to putting together the equivalent of a Broadway show every week! The credit on the show's crawl always read,

"Special Material By . . ." because that told the viewer that something creative was going on.

I was that guy on *The Dean Martin Show*. Plus, I had the added job of standing in for Dean all week long for rehearsals because he didn't show up until the very last minute before taping.

This is the story of what happened backstage and behind the scenes in those joyful years when Dean was such a success. It is also the story of three distinctively different men who worked closely together in the production of that show.

Of course, there was Dean Martin, a performer whose happy-go-lucky, uncomplicated public image camouflaged a personality that was extremely complex and private.

Also, any story dealing with Dean's enormous TV success must also focus on the unique man who so firmly guided that success, producer-director Greg Garrison. In their fifteen-year association, based solely on a handshake, Greg made himself rich and Dean richer. He also became something of a legend in the business. Talk to ten people and you'll no doubt get ten contradictory adjectives describing Greg, and there'll be an element of truth in all of them.

Greg Garrison's television résumé is long

and impressive. Just for starters, he directed *The Milton Berle Show*, *The Kate Smith Show*, some major TV specials, and spent some exciting times at the helm of *Your Show Of Shows*. Greg was the kind of guy who insisted on the best from his staff and performers. Once just before a Dean Martin roast, he pushed an inebriated Jackie Gleason up against a wall and leveled him with a forearm into his chest. That got Jackie's attention. He knew he was going to have to do the show (he was the honoree) or Greg would kill him. But Greg never gave him a shot to the face (he didn't want to ruin Jackie's makeup).

Lastly, this is partly my story — a boy from Tacoma, Washington, who on rainy days in the northwest used to dream of one day writing "The Great American Musical Comedy." I didn't do that, but I did spend fifteen happy, creative, and exciting years working for Dean and Greg with some of the greatest talents in the business.

During those years, I *was* Dean Martin, standing in for him on the show, writing for him, constructing his musical material, and *being* him during rehearsals. I sang with Kris Kristofferson, Tony Bennett, Peggy Lee, and Olivia Newton-John, danced with Cyd Charisse and Ginger Rogers — that is, until

Dean was ready to take his place in front of the lights and cameras. I worked out material with Dom DeLuise and George Burns, had Linda Ronstadt walk out on me in a snit, got told off by Debbie Reynolds, and even drove Raquel Welch to go see Greg one day to get something off her ample chest — in fact, she wanted me fired!

But I stayed and remained that starstruck kid, spending many interesting, dazzling, and rewarding days. I learned that Jimmy Stewart *was* as warm and personable as he seemed, Lucille Ball really *did* know what she was doing every minute, and Frank Sinatra *could* charm the daylights out of you. I also learned that Bing Crosby was a miser at heart, Kate Smith paraded around the studio like a dowdy housewife, and John Wayne bowed off the show because he was afraid a minor indiscretion of his might become public.

There was a passing parade of fabulous faces and tremendous talents — Gene Kelly, Bob Hope, Johnny Carson, Sammy Davis Jr., Lena Horne, and many more. To me it was mind-boggling.

Looking back, those were groundbreaking days in television. We worked hard, we had fun. We stretched for and often reached the heights. Most important, I believe we gave a great amount of pleasure to the public who treated us so well. I found it all fascinating. If at times the following pages seem like constant name-dropping, well, that's the way it was.

My days with Dean and Greg were fast-paced, inventive, and always exhilarating. We were high rollers and the stakes were extraordinary. This is a story of success and fame, of life among the good guys and the bad guys, and the wildly wacky assortment in between, all intermixed with their lovers, agents, and egos. It's a story about Hollywood.

Lee Hale

gave a great amount of pleasure to the public who frequented it ... as so well it round it all

1 The Beginning

Dean Martin and Jerry Lewis split up their hugely successful act in 1956. The breakup had been coming for some time. Maybe Jerry wanted too much control and maybe Dean was tired of playing second fiddle. Or maybe the two of them simply grew tired of doing the same old antics over and over. One day, for reasons even he couldn't remember, Dean walked away.

The two of them immediately began to build separate careers, although neither was nearly as successful in their solo movies as they had been as a team. The public was sure it was Jerry who would zoom ahead on his own — after all, wasn't he the funny one? And then there was Dean's first movie without Jerry, an unqualified bomb called *Ten Thousand Bedrooms*. But Dean had discovered an image — working in nightclubs. He was the "People's Drunk." Dean came out on stage with a drink in his hand and an off-the-cuff casualness that stated clearly to everyone in the room he was "just one of the guys."

And then came *The Young Lions*, a 1958

Greg "pointing the Italian to his mark"

film that changed Hollywood's thinking about Dean. He almost didn't get the part. Although he was in the running, studio heads insisted that Tony Randall was a much bigger name and a better actor. After Dean's first movie ventures without Jerry, he was poison to exhibitors. But costars Marlon Brando and Montgomery Clift felt Dean was perfect for the part and went so far as to refuse to show up at the studio unless they dumped Randall and hired Dean. They spent several days "on strike," playing golf with Dean until the studio relented. Dean got the part and nearly stole the picture from Brando and Clift.

Other good roles followed, as well as several hit records, and in 1965 NBC started negotiations with Dean's lawyers to add him to its list of variety series.

A weekly show? "No thanks," Dean said. He disliked the idea so much that when the network suggested it, he presented them with a list of demands he thought would be impossible to fill. He asked for an outrageous amount of money, of course, but there was more. He only wanted to work one day a week, and that day had to be Sunday. He didn't want to do anything but announce the guest acts. He didn't even want to sing if he didn't feel like it.

Maybe he realized that many of his friends had tried weekly TV and failed. Was he insecure enough to feel that he, too, might fail — again? It wasn't Martin and Lewis anymore. Just Dean. Alone. And maybe to television audiences he was still just another Italian singer.

But surprisingly NBC agreed to each of his demands.

"They should have thrown them in my face," Dean said later, "but they agreed to it all. So what the hell, I had to show up!" Sticking to all his rules, Dean showed up on a Sunday, and NBC, sensing possible disaster, booked a long list of top names to pop in and out. Old buddy Frank Sinatra was the main guest. The two of them and

Diahann Carroll ran through a medley about nothing in particular at the end of the show, Dean's only contribution to "variety." The ratings were pretty good for that premiere show, but subsequent episodes sent *The Dean Martin Show* plummeting.

The choreographer for the new little flop was Kevin Carlisle, with whom I had worked on New York television and for whom I was now feeling a little sorry. He had left what seemed like a flashy career in the Big Apple to go out to the coast and get involved with this turkey. After four or five shows, he called me in Greenwich Village and asked if I'd be interested in coming out and creating some musical material for Dean and his guests. The show was in trouble, he said, and director Greg Garrison thought that involving Dean more might cure all its ills. They needed a "music man" with ideas. He'd told Greg about me. Would I come out for a trial period? At my own expense, of course. Call it a vacation in Hollywood, if you like. He made it sound as though I was the only person in the world who could save the show from cancellation. Flattery got me to Burbank and directly into chaos.

"They just fired the producer," was Kevin's first announcement to me as I got off the plane. "They're letting Greg produce and direct. He's got some great ideas. I hope you have some, too."

I took a deep breath and walked into Studio 4 at NBC Burbank, in search of the two strangers I had great qualms about meeting.

Who was this Greg Garrison anyway? As a matter of fact, who was Dean Martin these days? A TV has-been.

I did know one thing about Dean that impressed me, however. In 1964, Dean's recording of "Everybody Loves Somebody" replaced the Beatles' "A Hard Day's Night" as number one on the national charts.

Little did I know then that these two men, so different from each other, would eventually shape my life and my career. Two such contrasting characters: one so stormy, the other so evasive. Greg was the strong, overwhelming boss and Dean the easy-going, self-proclaimed swinger. At the same time, there were some qualities they both shared. Dynamic, yes. Difficult, sometimes. Interesting, always.

However, on that day, as I walked down those long, cold, barnlike NBC halls, I felt I was about to walk into the principal's office.

Studio 4 is the "biggie" at NBC. It's the best audience studio because of its setup, the site of most of the network's big variety shows. *The Dean Martin Show* had it for two days a week, Saturday and Sunday.

I sat alone in the audience section; no one else was allowed there until dress rehearsal. Coming up at the moment was camera blocking, a time to set up all the camera angles, the lighting, the microphones; a boring couple of hours including all the technicians, guest stars, chorus, dancers — everybody but the star. He wouldn't be arriving until just before dress. Part of the contract, you know.

Before me was a scene of bustling activity. The great TV lights concentrated on center stage, where there was considerable last-minute hammering. Stagehands were putting together some colorful platforms that suggested a dance number of some sort was about to be blocked.

On the left side was a small set, warmly decorated to suggest someone's den. I was to learn that this area was known on the show as Dean's "home base." It was where he came to sing his opening song and read the cue cards for his monologue. Most of his guest introductions would come from there, too.

Over on the right was another small area with bookshelves, a grand piano, and an

24

earthy-colored couch. These side areas were comparatively dark at the moment. They obviously belonged to Dean.

The headliner for the week was Ethel Merman. On the bill were comedian Jack Carter, the New Christy Minstrels, a juggling act called The Carlssons, and two young singers, Joey Heatherton and Leslie Uggams.

Although there seemed to be a lot of hustling and bustling, voices on stage were almost muted, as though these people were an army of well-programmed robots in a science fiction world of tomorrow. NBC Burbank was new then, modern and sterile compared to the semirestored theaters the networks had taken over in New York. To me this massive sound stage seemed unreal, but still there was a tingling of excitement — as though something might happen any minute and nobody in the building was quite sure what it would be.

The feeling grew and grew as they approached the first piece of business. Without warning, the whole place shook with the playback of Joey Heatherton's production number, "Looking for a Boy." Joey barely bothered to lip-sync it, preferring to concern herself with the dance steps. The ear-blasting sound was extremely impressive to me — very contemporary, a nice change after all that Broadway stuff I was used to. A sharp contemporary approach to music, instead of the brassy screaming of musical comedy. This was fresher, with the slightly rock-and-roll background of the '60s. The New York musical comedy approach seemed old-fashioned by comparison. I knew I was going to like Les Brown and his Band of Renown.

Unlike the shows I'd done in New York, there was very little stop and go here. Straight through once or twice, followed by a loud voice from the control room. "That's fine. Next."

That's it? Like an assembly line, Leslie's two songs were quickly set and blocked, followed by the New Christy Minstrels, on and off. But where was the star?

Again, the voice from the control room. "Just show Jack Carter where his mark is. We don't have to go through his routine." It was Ethel's turn. She walked through her two songs, was told which camera to look into at what time, then, "Okay, take ten!"

Still no sign of Dean. Finally, after everybody else was taken care of, he materialized. He was every inch the Hollywood star. A beige alpaca Sy Devor sweater, very expensive-looking and well-fitting pants, beige

Greg Garrison
holding Dean's cue
card

shoes. He gazed around as though he wished he were somewhere else — anywhere else. The dark eyes moved, but they seemed to focus on nothing. His tall, slim, muscular form barely moved.

Standing just an elbow away was another macho man in a bright red shirt and faded jeans. Must be Greg, I thought. Handsome guy. Younger than I figured. Looked a bit like Cary Grant on one of his off days.

He had his arm around Dean's shoulder. He was talking to him almost in a whisper, confidentially, while at the same time slightly pushing Dean toward center stage, maybe into a better light. Both of them played "Mr. Cool" to the hilt. They seemed to be locked together — like Siamese twins.

Where one went, the other went as well.

He took Dean over to Ethel Merman and introduced them. You mean, it was almost time for dress rehearsal and this was their first meeting? Incredibly, yes.

Dean started to rehearse a medley with Ethel, a collection of songs she was associated with. Dean seemed uncomfortable in his sections. He stopped at each entrance and tried again. I sensed trouble. Quietly, Greg whispered something in Ethel's ear and they started again, Ethel singing some of the parts assigned to Dean. Once they got to the end, Ethel walked off as though she was glad it was over with, and Dean was hurried over to his couch in the "piano room" set. He was obviously more comfortable there, chatting freely with the cameramen and the stage crew, joking about "that Broadway medley" with Ethel.

"Boy, has she got a great pair of lungs!"

He kidded with accompanist Ken Lane. "Ken's been with me for seven years and has had a thousand offers to leave. I know because I made them." Dean sang a parody:

> *"It's quarter to three*
> *There's no one in the place except*
> *You and me . . .*
> *So stick 'em up, Joe!"*

28

Dean and Kate Smith

Everybody laughed. "Now let's get serious," Dean said. He and Ken, the choir, and Les Brown's band did a chorus and a half of "Clinging Vine." Once through, Greg pulled Dean off the couch and told him he could go back to his dressing room. "Can this kid sing?" Greg yelled as Dean wandered across the stage and out of sight.

"You're damn right! Did you hear that, Ethel?" Dean called out. He returned to the cocoon of his dressing room and Greg went back to the control room, where he ran everything by booming out orders over the P.A. system.

An hour later, dress rehearsal. Everybody

in the cast was in "dress" but Dean. Despite the fact that there was a studio audience, he was still in his beige alpaca sweater.

I sat and wondered what part I could play in this little drama. What about Dean? He seemed so uncertain. Would I be able to rehearse him? And what about Greg? Was he really that tough?

Maybe I should forget the whole thing, get on a plane, and fly back to comfortable old New York. There was time. I still hadn't met anyone. I was terrified yet intrigued.

Right on schedule, the taping brought in a whole new audience. I wondered if decisions had been made between dress and now. Would one audience influence the jokes or the music? I'd soon see. The Voice from above quieted the audience and introduced Dean. He was greeted by a tumultuous round of applause and cheers. He managed a couple of clean jokes for the crowd from Peoria, and Greg sent him off to his starting mark. There were a few stops here and there to change scenery, but they were brief. Mostly, it was gung ho, straight ahead, as though the show really were live. It was scheduled for an hour from seven to eight p.m., but took a little less than that. "We'll spare you the commercials," the Voice said.

The show was a professional but rather dull parade of vaudeville acts. The best moments were those between takes when Dean had a little fun with the audience, or flubbed an intro. The audience liked that. Actually, I did, too.

The Dean-Ethel medley went along nicely. The audience applauded on cue whenever they recognized a Merman show-stopper. Dean was polite but laid-back. Not until he got over to the piano with Ken Lane did he seem to wake up. The silly parodies, the ad-lib bits with Ken, the well-sung ballad. Suddenly the whole studio was alive! I made a mental note.

The show's offices were two blocks west of NBC in a one-story L-shaped building on the corner of Hollywood Way and Riverside Drive. Greg and executive producer Hal Kemp had offices back in the long end of the "L," with my friend Kevin Carlisle at the short end and the reception desk smack in the center between the two.

The lady at the desk asked my name and business, which I quickly said was to see Kevin, leaving what I thought would be an intimidating meeting with Greg Garrison until later. She buzzed Kevin, but before he appeared there was an eruption of shouting coming from the end of the long hall — something I would later learn went with the territory.

From Kevin's advance publicity I knew the loud discussions must be from Greg's office. I thought seriously of turning around and making my exit. Even though I'd arrived feeling secure about my background in music and television, I suddenly had the uneasy feeling that anything I suggested would be denounced loudly and furiously. All of Burbank would hear. No, better for me to return to the sanity of New York. I was actually heading toward the outside door when Kevin grabbed me and dragged me down the hall to his office.

"Pay no attention to that, nothing to do with you," Kevin assured me.

Maybe. Not yet.

"See this hole in the plaster?" Kevin pointed to a huge crater in the wall outside his office. "Sometimes the only way he can let off steam is to hit something." Where was that exit door?

Kevin pulled me back into the office and asked his assistant, Dru Davis, to get us some coffee. He calmly tried to explain that the show was not going well and Greg felt the only way it could be saved was by involving Dean more. "That's why you're here. Let's go see if Greg is busy."

He literally shoved me down the short end of the L. Midway, at the reception desk, we heard Greg's voice bellow out some obscenities that seemed to faze no one in the immediate area, and Kevin kept pushing. Door, where the hell was that outside door?

But Greg suddenly appeared before me, exiting a writer's office in the middle of a "son of a bitch," and smiled broadly. A tall, handsome hulk of a man, he looked more like a defensive end for the Green Bay Packers than a TV producer.

"Ah-hah!" His energy switched gears and a loud but obviously sincere fuss was made over my presence.

"Our boy from New York. Good, good! Come on down the hall — we'll talk."

Greg's office was a small anteroom off Mr. Kemp's lavish suite. It had just four pieces of furniture, a round neon blue table that looked as though it had been dragged in from the office kitchen and three folding chairs. The only thing on the wall was a large 1965 calendar with some checks and Xs and last names of stars written in red over various dates. Maybe there was a purpose in the "desk" being round. Guests might be uneasy, asking themselves, "What's going on here?" But to this man, who was then up against the wall doing a series of deep-knee bends to "get the kinks out," this was where everything was under control. Greg didn't need a desk.

It was also obvious that he spent little time in this "office," preferring to wander down the halls, chatting with the writers, telling raunchy jokes, screaming about those "fucking agents and managers," generally letting off steam, and picking up his phone calls at whatever office he happened to be passing at the time.

Mickey Rooney, Barbara Eden, Dean, Kate Smith, and Norm Crosby

He asked me to sit. Like a scared puppy, I sat.

He chose to stand in what might be called

the "Garrison Stance," feet shoulder-width apart, planted firmly on the ground, arms folded. That day he was wearing another bright red shirt and another pair of jeans held up by a thick belt with a huge buckle. Macho all the way.

He took some time to look me over, checking me out in his mind as though I were a new horse in the stable. When he spoke he avoided the business at hand and with glowing homilies compared the wonders of Southern California to the dirt and grime and foul weather of New York.

"Well, now, what did you think of the taping?" he asked. I mumbled something about Dean making Perry Como look active. I was hoping he would find that funny. After all, Perry at that time was known to be the epitome of calm and relaxation. Greg chuckled just a little and let some time go by before his next statement, giving it extraordinary importance. This was meant to stick: "Dean knows one thing very well, and better than anyone else. He knows how to be Dean Martin."

Another pause to let that sink in. The silence was interrupted by a series of phone calls, all of which he took, carrying on what seemed to be private and crucial conversations. Was I being tested? Would he check later to see if I had betrayed him? My eyes traveled to the calendar. Although those phone conversations did seem interesting and terribly important, decisions about well-known stars were made on the spot, and there wasn't the slightest attempt to keep them from me.

I felt Greg was telling me, in his own way, that he approved of me already and that I was now part of the family. Actually I did very little talking that day. He did it all. Not much was said about the show, my background, or what was expected of me. Kevin had given him a glowing report about my credentials and he obviously was ready to let me prove myself.

Dean and Greg

At the next break between phone calls he insisted on telling me the latest joke from the writers' room, followed curtly by "Why don't you go down the hall and let Kevin tell you what we need here. I've got a couple of calls to make."

After I was out of earshot, although I was never *absolutely* sure of that, I said to Kevin, "He's quite a guy! I think I like him!" I was strangely fascinated by it all. Even though it was the oddest meeting I'd ever had, with hardly a word about my future in this office, I felt I'd done okay, showing neither fear nor shock. That seemed to be important to him at the moment. The rest of the matter would be decided after I submitted my first material. Now it was up to me.

I was given a long list of all the guests who were to appear on the show for the rest of the season — seven to ten acts per show, all booked right up to the last week. Executive producer Hal Kemp had been a well-respected talent booker for years. But a run through these pages told me that even though he had contracts signed well in advance, he had not tried to make interesting combinations of stars, something that would improve the show. He'd just made a list, whoever was available.

Greg made no secret of the fact that he

considered the show highly overbooked, and that if he had his way he would cancel all the novelty acts and stick with the stars. With Kevin his only accomplice prior to my arrival, he had now started his master plan to get Dean appearing more and enjoying it more without realizing there was any change.

Greg came bustling into Kevin's office on one of his journeys down the hall. This time he sat down.

"This will be the longest meeting we'll ever have. Pick two guests from this week's list that you think Dean could work with and let me see what you can do." I nodded. He was up on his feet and out the door. I watched him disappear into head writer Paul Keyes's office, probably to discuss what they'd do if I botched it up.

I gazed at that week's list: Tammy Grimes, Mickey Rooney, the Amin Brothers, Jane Kean, the Krofft Puppets, dancer Elaine Dunn, and Kate Smith. Kevin said he had already planned production numbers with Tammy, Elaine, and Mickey that I could help him with, but my eyes set on the "biggie," Kate Smith.

It was so simple: a medley of tunes both Dean and Kate were associated with, only she singing his songs and he singing hers. Because of the trouble I'd witnessed during the rehearsal of the Ethel Merman medley, I wrote in the choir to sing lots of melody in case Dean got lost. Nobody could miss that, I thought.

I typed out all the lyrics and handed them to Greg to look over on one of his jaunts down the hall.

"What took you so long?" he joked. It had only been an hour since our meeting. "Perfect! Who else do you want to do something with?"

I suggested Mickey Rooney as the obvious "other biggie" and he agreed. I asked Ken Lane to jot down any songs he felt

Dean at Ken Lane's piano

Dean would know but hadn't done in a long time. He came up with five or six. Somehow I made sense out of a medley for Dean and Mickey that included "The Boy Next Door," "I Don't Know Why I Love You Like I Do," "The Girl That I Marry," "Once in a While," and "Nevertheless." Greg was very impressed and called me in for another meeting.

"This will be shorter than the last one. Move your things out from New York. You've got a permanent home in this office." He yelled out to his secretary. "Joan, get all that junk out of that little room next to Kevin's and . . ."

His voice trailed down the hall as I felt my life being turned around by this Greg Garrison.

I ran after him. "But I've only been here a few hours. Doesn't Dean have to hear these things?"

"If I like it, he'll like it."

"But it's all so simple."

"What's wrong with simple?"

So far all I'd done was write a couple of medleys and hang around waiting to see the stars who were going to perform them. And it was only my first day.

The next morning the music staff (con-

40

ductor Les Brown, arranger J. Hill, and Ken Lane) dropped by to scrutinize the work of the new boy and pass their judgment on to Greg, and, I assumed, to Dean. Uninvited, I listened from down the hall as they reviewed the Kate Smith and Mickey Rooney routines. They ran through them once and came to the conclusion that what I'd done was indeed passable. "Dean won't have any problem with these" was their report. It turned out to be the last such organized meeting. From then on I would simply sketch out my work and call in J. Hill when we were ready to orchestrate. Les Brown said he felt comfortable that all would go well and that he'd much rather be out on the golf course anyway.

There was one other regular meeting that became obsolete after that week. Previous to my arrival, Greg, Les, and Ken Lane would go to Dean's house to run through the music and sketches, something Dean frowned on from the beginning. Now Greg started to implement his master plan: Don't tell Dean anything! Keep him away from all rehearsals, let him show up at the last moment and be surprised. He wouldn't even be told who the guests were. He'd meet them when he got to the studio. The less he knew about what we were doing the more spontaneous the show would be.

Getting down to rehearsal, Kate Smith, as all guests would, kept asking when Dean would be there to sing with her.

Trying not to frighten her, I explained that Greg felt the show would take on a new and original look if we all did our homework and let Dean wing it when the cameras rolled.

"I've adored that man for years," she confessed about Greg. "And if Greggie thinks Dean can pull it off, then we'll all try to

Dean singing his weekly ballad

make him look good." "Greggie" had steered Kate's long series of one-hour daily variety shows from NBC in New York. She gave him the lion's share of credit for their success and was behind him in this strange new television adventure — no rehearsals.

But everyone else would have to know exactly what they were doing as well as be prepared for disaster if Dean should go blank. As the TV audiences would learn, that never happened. On the contrary, Dean always came through.

He did ask for one small rehearsal, however, before show time. He wanted to hear the orchestrations. So each week we scheduled a band rehearsal just for him and any of the guests with whom he would be working.

"Just let me hear it once, and I'll be fine," he insisted.

Dean's first call was 1 p.m. Sunday. The show would be taped at seven. In between would be camera blocking and dress rehearsal.

In both cases, Dean would simply watch me go through his paces on a monitor in his dressing room, hopefully paying attention enough to be aware of things when taping began.

Dean was seldom, if ever, late. He'd arrive promptly at one and go right to the or-

Dean

chestra pit to run down his music. It was also the first time he'd learn who his guests were, and since Greg didn't want too much socializing before the taping, I would rush through the introductions and the explanations as fast as possible.

As I stood by the band, waiting and hoping that my material would be accepted, I feared the worst. Even though Kate and Mickey had learned their parts, none of us knew how Dean would react. And here we were a few hours away from taping, orchestrations already on the stand, cameras positioned for the moment of truth.

Dean sailed through the door with quiet

abandon, wearing an alpaca sweater similar to the one I'd seen him in the previous week, only this one was yellow. So were his socks. His brown hair was tousled, his eyes clear, his air flamboyant.

"Hi, guys, is this where the action is?" Big laugh. Les's band obviously liked Dean. He had an appreciative audience for a couple of off-color jokes, whispering the punch lines almost as asides so that the one female member, the harpist, might not hear. She did and enjoyed the story as much as everyone else.

Greg pulled me over to introduce me to Dean as the laughs subsided. I didn't know it at the time, but I was one of the chosen few to have that privilege. Most of the staff and crew were kept away from the star.

Greg told Dean that I would be writing some of the musical material for the show.

"Hi, kid." A nice, warm handshake, but Dean's eyes roamed almost immediately toward the cue cards.

"Let's see what this one's all about," he said, as though our introduction was holding things up.

Greg had instructed me to stand close by Dean during these orchestra readings to help him with his entrances.

To my relief, he didn't seem to need any

help. Each time I opened my mouth to sing along, he was already into it himself. I shut up.

Just as I moved back, feeling totally unnecessary, he turned to me and quietly asked, "How does that go again?" Les started to wave the band to stop playing.

"Don't stop," Dean said quickly. "I'll get it." As we rolled on, I sang the troublesome spot into his ear softly. Next time it came along he breezed through it with no trouble at all. This was going to be our pattern for the next fifteen years. At the end of the song, Greg pulled me aside. "You see," he said smiling, "simple."

Dean was *comfortable*. That was the key word. Make Dean as worry-free as possible. It had to appear to everyone that this was *easy* for him, and it was up to me to make it that way.

Watching this pre-show performance of "See How Easy Dean Does It," it occurred to me how alike the three of us were physically. All about six feet tall, dark and slender, Greg and I a hair younger than Dean and maybe not *quite* as handsome — but then *he* was the movie star!

There we were — Greg helping Dean with asides to the band, Dean delivering one-liners to Kate Smith and Mickey Rooney and making it look like it was all just a lark. Dean seemed not the slightest bit concerned that he was just about to stand in front of millions of viewers, while I waited silently in the background, hoping my creations would somehow help bring it all together.

Whatever would happen in front of the cameras in just a few minutes was anybody's guess. But the basics were in place and apparently working. It did work, and I immediately thought of asking for more money. After all, I was getting a pittance. I hadn't accounted for the miserly ways of Hal Kemp, with whom I had had a conversation as long as the one with Dean.

"We'll get to the old boy later," Greg kept saying. Mr. Kemp (as he was always called) was in his eighties, had a long and impressive booking career, and was a hard man with a dollar. He'd already spent beyond his limit on guests for the show, where he felt most of the money should go. That is, what was left after Dean's big share came off the top.

Mr. Kemp had no idea what I was doing there, other than occasionally singing with the off-camera singers, so any change in the weekly budget to accommodate me had to be handled very carefully. He was a good booker. Period. The rest of the machinery of putting on a show could be put in someone else's hands — fortunately for Mr. Kemp, these hands were Greg Garrison's.

The Dean Martin Show was in trouble because it looked like just another succession of vaudeville acts with an occasional star performer. Dean's contributions were minimal, but Greg was determined to change that, regardless of what his star and the network had agreed to. Now that the original producer had been fired and Greg had assumed one of television's first hyphenated positions as producer-director, he had the opportunity to get Dean involved more while still showing up for the same number of hours at the studio.

Dean was always impressed with Greg's maneuvers to make his working day shorter. "He not only sticks to the damn schedule," he told the network heads, "but he usually gets me out ahead of time" — sometimes, unfortunately, at the expense of some of the guests.

"There's only one star around here," Greg always told us all. "If the guests don't like it, tell 'em to get booked on some other show!"

He knew from the beginning, of course, that it wouldn't be easy. He confided to me that he needed help getting Dean to work with the important guest stars on the show and still enjoy himself, still *be* himself. Greg's master plan was under way.

2 The First Year – (1965–1966)

The star of the show I was now working on had said only two words to me at our introduction: "Hi, kid." It looked as though it was going to be nothing but guesswork trying to figure out what songs he'd know and what songs he wouldn't know, never being able to go over them with him, never even being able to discuss them with him. Was I never to engage Dean Martin in any kind of conversation? Was I, in fact, dealing with a phantom star?

If I was going to live with all these restrictions, there were a few things I had to do. From Neil Daniels, the president of the new Dean Martin Fan Center, I got a list of every record Dean had made. There were a lot, and I figured he might remember some songs from them. Since I could rehearse the guests, whether it was Peggy Lee or Carol Channing, I'd have them begin each section so that Dean would hear eight familiar bars and hopefully glide along from there. Sort of the blonde leading the blind. Because Dean and I were more or less contemporaries, I figured he'd know the same songs I knew. If I followed some good sense and intuition (with a healthy load of luck), maybe he *could* walk in at the last minute and sail through all his musical bits without the benefit of rehearsal. It was to be a game, a big and expensive one, too: Television Variety Show 101.

The second week brought the second big challenge, another of Mr. Kemp's overbookings: Milton Berle, Imogene Coca, the Supremes, Jane Morgan, the Step Brothers, Joey Forman, Jackie Mason, the Krofft Puppets, and Herb Alpert and the Tijuana Brass. They were all scheduled for the same one-hour show. There didn't seem to be room for Dean, much less time for him to do anything with anybody. I decided that for the first time we'd do an out-and-out finale using everyone. Maybe it would make it look like a real variety show for a change instead of just a parade of acts. Greg Garrison was thrilled — two appearances for the price of one.

Greg still wanted to make room for something special with Milton. Maybe Imogene, too. I had thoughts of Milton in drag marrying off his daughter, Imogene, to Dean, possibly even throwing in the Krofft Puppets as their offspring.

Patti Page and Dean

"I'm just happy to be here," Imogene kept assuring me, insisting she'd do anything we asked her to do. Sweet lady.

I took up the matter with our choreographer Kevin Carlisle. "Coca's always had a marvelous comedy-tragedy feeling about her," I said to him. "Let's do a clown number. You know, the old chestnut about a lonely kid who's only happy when she's in makeup because that's when everyone smiles at her."

We came up with what we thought was the perfect song for it, *"People, people who need people . . . are the luckiest people in the world."* Imogene loved the idea but kept asking that same old question, "What am I going to sing with Dean?" Like so many other guests, the most important time was the three or four minutes with our star, which, of course, was more than I had with him.

We devised a way to bring Dean in at the end of her number as the only person at the party who pays any attention to her (and the only one wearing a tuxedo). The obvious song for Dean: *"When you're smiling, the whole world smiles with you."* Imogene was in seventh heaven. We thought she was going to faint when Dean approached her during the opening bars. But then, as was the case with most guests, this was their only moment together. Dean would be hers once and only once.

During the rehearsals I stood in for Dean. It was the beginning of what may be a one-of-a-kind showbiz job. All week long I would be Dean Martin while the real thing was off somewhere playing golf, making a movie, or simply raising a few quaffs. Suddenly I was selecting the guests' music, writing their material, directing them and showing them how to react when their missing costar finally did appear on the fateful taping day.

At first I simply used my instincts — what I thought Dean might do, how he'd carry on with his guests. As the weeks went by I had a chance to study him more carefully, and I became more and more accurate in my judgment of just how he would behave.

Greeting one of our guest stars for the first time, I'd introduce myself and explain that I'd be going through the motions as Dean. I'd try to tell them what to expect, often giving them several possibilities. I'd prepare them for Dean's timing, predict his sense of humor, and tell them where he'd be sure of himself and where he'd need help.

No imitations of Dean, no singing impressions, no wisecracks, no alpaca sweaters. Just the basics. I'd try to comfort the guests, and made sure they knew in advance that Dean was extremely personable, exceedingly kind, and always the gentleman.

I learned to move, think, and feel like Dean — kissing and hugging all the lady guests and trading dirty stories with the guys.

Most guests graciously put themselves in my hands, but there were some who not so quietly resented the fact that they couldn't work with their costar until the last minute. No matter how I tried to convince them that everything would turn out just fine, they were often so nervous because of it that they felt they had every right to explode somewhere along the way. More about those guests later.

Milton Berle had his own idea about what he would do if Greg insisted on a musical turn for him alone. He had always wanted to play the lead in *The Music Man*.

"Nobody ever asked me," he said. The big song from that score, of course, is "Ya Got Trouble," one of the toughest five minutes in the Broadway catalogue. Milton considered it a marvelous challenge. He just knew he could handle it.

We rehearsed it over and over again, exhausting three rehearsal pianists, who took turns. But that song's a tough go and you just have to sort of plow through. We certainly did that. But even after endless runthroughs, Milton was still not secure when taping day came.

"I'll get it, I'll get it. Just do that one part over again for me." That meant "from the top" and another round of musical chairs for the pianists.

Behind all that brashness, Mr. Television was surprisingly insecure. Was it always like this or was it a result of lost glory? Was he finding it more and more difficult to recapture those days when he was the king?

Greg had directed some of those early shows but didn't prepare me for Milton's overwhelming ego. He needed constant reassurance. We were all exasperated that anyone would take so long to learn that song. But when our patience was gone, Milton brought in his own piano player, Hal Collins.

"Terrific!" Hal would say after each Berle attempt, good or bad. We finally realized that this was what Milton needed, so we went along with it, praising each repetition no matter how miserable it was.

Greg got right to the point: "If this guy needs to be inflated, let's blow a little smoke up his ass."

Milton certainly did have drive. But I wondered how many other star egos I'd have to coddle in this new job.

Dean was just the opposite of Milton. Nothing seemed to be a problem. He could carry off a mistake with great humor, usually so well that everybody assumed that was the way it was supposed to be in the first place.

Milton wanted to do his best for Dean, for Greg, and most importantly, for Milton. But it seemed he was deliberately torment-ing himself by having attempted such a diffi-cult piece of music.

Finally I devised a way to do it in sections, hoping Greg would buy it. Milton would only have to worry about one minute at a time. But Greg felt the song would only work if performed straight through without stopping.

"You son of a bitch! You're trying to ruin me!"

"Milton, when you did *Texaco Star Theatre* you didn't stop tape, did you?"

"That was a live show!"

"This is gonna look live!"

Back to the piano. One more day and Milton might have made it. He tried very hard, but during the taping he bogged down at three different spots. Fortunately they were right where Hal Collins and I had figured the stops earlier.

Sid Caesar and Dean

Milton was embarrassed in front of our live audience, but Greg told them it was due to "technical difficulties" and later edited it together to make it *look* live.

Things didn't go so smoothly with the Supremes. The girls themselves never said anything about it all week, although the young and very mousy Diana Ross seemed terrified of everything. She sat with me in the bleachers, begging me to tell her how to act with Dean. I tried to calm her down by insisting that all she had to do was be herself and think of Dean as a warm and gentle friend. "That's how comfortable he'll be with you," I assured her.

But there was another problem. Excited as the Supremes were about working with Dean, Greg suddenly nixed the medley I'd planned for them.

"Just let them sing their solo and the finale," he said. "That's all they're worth to me!"

The girls were disappointed and immediately complained to their managers, who were always hanging around in the shadows like vultures, ready to pounce upon anyone who didn't treat their clients right. They immediately rushed to Greg in their no-nonsense way and loudly warned him that if the girls didn't do a song with Dean, they'd walk.

Barbara Itzikman (one of our dancers), Nanette Fabray, and Dean

At first Greg attempted to politely explain that the show was overbooked and there simply wasn't enough time to do what they asked. Tempers began to reach the boiling point and, sensing the crew and audience reaction to all this, Greg quietly escorted everybody involved off the stage and into the hallway.

Diana, Mary, and Florence, however, remained with me, smiling as best they could, hoping that the little scene would resolve itself and everybody would be happy.

It wasn't long before one of their managers appeared and gestured them to follow him out the door. The girls responded in an instant, leaving me, the crew, and of course the audience hushed, waiting for the outcome.

I was close enough to the hall door to hear Greg shout, "Nobody tells me how to run my show! If you don't like the way it's set up, then let's forget the whole thing."

Suddenly he rushed through the studio door and smiled broadly as he told the audience that we were ready to begin the taping. He gave no indication as to what had happened.

The Motown entourage, including the Supremes, had skittered away in the opposite direction, spouting various raunchy opinions of Greg and his show and what he could do with it.

I frantically rounded up Imogene Coca and Jane Morgan, telling them that they'd have to sing the Supremes' sections of the finale and that I'd go through it with them when I had time.

Up in the control room, we all wondered what we would do when we came to the girls' solo spot.

"Skip it," Greg muttered. "We're not holding up the show for them."

But just as the rundown came to the girls' number, our stage manager called the control room to say that they were standing

Dean and Carol Lawrence

backstage next to him, ready to go on.

"Push 'em out and let's go then," Greg grumbled. No one was ever sure what exactly changed their minds. Perhaps a bit of "the show must go on." But they did a bang-up job with their number and no further mention was made of their spot with Dean. To my relief, they also insisted that after they changed their dresses they'd be out for the finale.

Greg walked down from the control room, whispered a few pleasant words to Diana and company and gave them each a little kiss. Everybody was all smiles. "Get the Italian out!" he yelled to the stage man-

61

ager, and it was on with the show.

Dean, as usual, had been hidden away in his dressing room and was totally unaware of any of this; all he knew was the score of the day's game — Oakland Raiders 21, New York Jets 20. Everything went as originally planned, but the Supremes never appeared on the show again. Greg Garrison doesn't forget!

An old friend topped our guest list the following week, Tony Bennett. He confessed that he was in total awe of Dean, and that he admired him because he had his own show and did it without the drudgery of rehearsals.

"I don't know how to react to all this," Bennett kept saying. I assured him that the way we had it set up, Dean would be comfortable with Tony musically and told him not to worry. But Tony balked at the fact that I had to take the keys down for Dean's vocal range. Eventually he agreed that, after all, it *was* Dean's show and Dean came first.

The nervousness showed in the beginning, but after they got started there was a camaraderie that thrilled us all, especially Tony. Greg was off-camera, hopping up and down as he sometimes felt he had to do to get the joint jumping. But it wasn't necessary. Tony enjoyed every minute.

Tony shook his head as he left the stage, but said he'd be very happy to come back again. "I can't believe he never rehearsed that medley!"

The following week the schedule gave us a break — not quite as many guests. There was time for Dean to work separately with every one of them.

"He didn't have to come in early and he didn't have to rehearse," Greg reminded everyone as the taping began.

With Canadian newcomer Rich Little, Dean did a "mimic" bit, then a short tune with the Krofft Puppets, half a chorus of "Side by Side" in English and French with Line Renaud, threw straight lines to Gene Baylos, performed a long medley with Louis Armstrong, another one with the Andrews Sisters, and a song with Carol Lawrence.

After Carol had finished her elaborate number with the dancers, Greg as usual wouldn't stop to let her catch her breath. He always wanted the show to look live, so lots of heavy breathing never bothered him.

There was some breathless patter between Dean and Carol that led into their duet, "Somebody Loves Me." Dean got to the lyric, *"for every girl who passes by. . . ."* but he was so incredibly far off from the melody that he knew he had to stop. There was no way he could charm his way out of that one. He stopped the orchestra with that coy smile and shouted over to Les Brown, "Somebody in the band must have hit a wrong note. Let's do it again, only no clinkers before I come in, okay?" Knowing giggles encircled the stage.

From the control room came Greg's booming voice. "We'll take it from your cross to Carol."

"Never say 'cross'," Dean yelled back, making the most of the situation. The audience was in hysterics and ate it up.

"Okay, we'll take it from your *walk* to Carol."

Greg cued applause again as if they were at the end of Carol's number. Dean and Carol rattled off the same dialogue and the same jokes that the studio audience had just heard, only this time there was deadly silence.

Dean took care of that. He looked squarely into the camera and explained to the viewers at home. "We did this all before, you see, and I made a mistake and *that's* why they're not laughing."

They began the song again until Dean broke in, "I have a right to make mistakes. It's my show!" Big laugh. When he got to the *"for every girl who passes by . . ."* section, he slid through rather nicely with Carol's subtle assistance. The audience cheered and applauded.

Dean was very surprised. "I can make more mistakes," he offered. We aired the *entire* episode, unaltered and uncut. Pure Dean.

I had worked as choral director on Carol Burnett's *The Entertainers* series on CBS in New York just before I came to Los Angeles. Carol flew out about the same time, even though she had just broken her leg in a backyard baseball game with new husband Joe Hamilton's kids. She had come to find a house for herself and Joe and the possibility of more children. Joe had to remain east to discuss with CBS her new weekly series, which he was producing.

I knew Carol would be intrigued at how we did our show without the star rehearsing, so I got her to venture out to Burbank to watch a taping.

She and my friend Ken Allen insisted on simply sitting in the studio audience and not making any fuss. Sometimes the audience didn't recognize her immediately, its only concern being how this girl with a broken leg would manage the steps. Against her wishes, one taping night I mentioned to Dean that she was there. He waited for a break, then told the audience that "this terrific lady with whom I've had the pleasure of making a movie" (*Who's Been Sleeping in My Bed?*) was sitting there among them. He introduced her, then ran up to where she and Ken were sitting and gave Carol a big, sloppy kiss. She responded by throwing her crutch down the steps, jumping up and sitting in Dean's arms, bringing a gasp from Ken and me and a roar from the audience.

Dean and Tony Bennett

"This is my kind of girl!" Dean yelled out. "I gotta get her on my show one of these days." He never did that, but then he had nothing to do with the show's bookings or really cared *who* was on each week. That was a part of Greg's job and he was not a fan of Carol's at the time.

To prove that she had a certain amount of clout in Tinseltown, Carol loved to take Ken and me to such grand eating palaces as La Scala and Chasen's, exclusive Hollywood eateries where *they* knew her even if Dean's bussed-in studio audiences didn't. We always got an "in" table and Carol loved saying hello to Rock or Tab or Cary. She was, and always will be, a movie fan.

"Gosh, there's Lana Turner!" The headwaiter would make sure we were introduced to any star Carol hadn't met yet. "Isn't it exciting?" she'd say. "They're all here!"

One evening at Chasen's we arrived a bit late and the place was packed. The maitre d' knew Carol by this time and escorted us into a lovely back office to wait our turn at a special table. Another gentleman was also waiting for a table — I saw that it was the famous Broadway composer Arthur Schwartz. We introduced ourselves, and he asked what line of business we were in. I told him that Carol was getting a new weekly TV series, and that I put the music together on *The Dean Martin Show*.

"My favorite show!" he shouted. "That guy is terrific. I watch it faithfully every Thursday." He wanted to know all about Dean, and if it was really true that he never rehearsed anything. "If he doesn't," said Arthur, "the man's a genius."

Carol found a house, a beautiful English Tudor mansion just off Sunset Boulevard near Doheny Drive, and although we all thought it was perfect for her expanding family, we suspected the real reason she chose it was that it had once belonged to Betty Grable and Harry James.

"If these walls could talk," she said as she led us through the master bedroom, "or sing or tap or play the trumpet. I think I can still hear them. . . ." She sat on the bed and began to sing.

"I can't begin to tell you how much you mean to me . . ."

I'd grown up with the music of the Andrews Sisters, so it was a great delight to find them as outgoing and fun as I'd hoped. Down to earth and very professional. I'd worked out a solo number for them based on Jerry Herman's song from *Mame*, "That's How Young I Feel." They heartily agreed with Greg's suggestion that they make their entrance on motorcycles. And why not? Three "Boogie Woogie Mamas" in long dresses riding sidesaddle on their Hogs. It was easy to throw together a medley for them and Dean that included a bunch of their hits. I knew Dean would have no trouble getting through any of them. But I asked the girls to hum along at all his entrances, just in case.

That medley also started us out on what was to become another Martin trademark: the pouf, a round overstuffed love seat. Dean didn't realize it was on a platform and as he was singing he tripped and stumbled onto the pouf. The girls made it all look like it was part of the act, and they had a good time. So did the viewers.

Louis Armstrong was working at Disneyland that week, so Greg suggested I drive down to his motel with the Dean-Louis medley on a cassette to work out any details. As with all the old pros Louis was "delighted to be on the show" and a medley of southern songs would be just fine. Since there were no problems at all, I said he could just show up on taping day and we'd wing it.

"Boy, your show really is relaxed," he noted.

Louis and Dean were a natural combination. Halfway through their medley Dean couldn't resist taking out his famous red handkerchief and wiping his face *a la* Louis, which broke up Satchmo.

We had, in fact, become so relaxed that production meetings were down to a minimum. Choreographer Kevin Carlisle and I simply talked over with Greg what we had in mind for the musical numbers and went ahead and did them. Greg sometimes dropped by rehearsal halls to see how things were going, or waited until Saturday when everybody got together for camera rehearsal. It was so laid-back that even with the major ingredient — Dean — missing, we knew it would somehow come out fine. Best of all, the ratings, which had plummeted after the first show of the season, began to rise significantly.

"We must be doing something right," Greg grinned.

Years of playing nightclubs with Jerry Lewis prepared Dean with seemingly endless supplies of one-liners to handle almost every possible situation — from Jerry's wild maniacal behavior to the antics of uncontrollable drunks. He survived all those years with a bountiful recall of zingers, putdowns, and pleasantries, plus an uncanny ability to make mistakes look like brilliant pieces of business while graciously allowing himself to be the butt of the joke.

Dean simply didn't need rehearsal. If he had known in advance what was happening it wouldn't have been half as much fun. There was that one and only rehearsal with the music that Dean insisted on. We set aside thirty minutes to run through it all, with the entire cast at Les Brown's orchestra stand. Dean would arrive precisely at 1 p.m. with his manager, Mort Viner, and we'd all be ready for him. It was his first knowledge of who was on the show and what he'd be singing. You'd think he'd be a little nervous, but he was always brilliant. It never took more than thirty minutes, usually less, because by that time the football games began and Dean didn't want to miss any of the action. In fact, we'd keep an eye on a TV set as we went about our business, waiting until halftime to get Dean on stage. It wasn't that we were really spoiling him; it was just an unspoken part of the deal.

Since I got no help from Greg, Mort, or anybody on the staff, I had to feel Dean out musically myself. I'd sometimes throw in high notes, just to see if he could handle them, and how far he could actually go, from one end of the scale to the other. It was sneaky, but it worked. If I feared he might have trouble with a tune, I'd put in a lot of melody that he couldn't help but hear. As for tempos, I'd start them out where I thought he'd like them, then let him take it faster or slower as he desired. If there was any hesitation, I'd be sure I was standing where he could see me without moving his head, then point to him just before his entrance in the song. All of this was worked out without a word being said between us.

Realizing he wanted it to appear to everyone that he really didn't need any help, I made my gestures more and more subtle. I understood that my anonymity was important to him and that he thoroughly enjoyed the game we were playing. Everybody watched with amazement. "How does he do it?" they would ask. "He really can wing it every week, can't he?"

I enjoyed the game, too. The less pointing and whispering I did, the better I'd done my job. Greg Garrison's master plan was in full swing.

"We just have to make sure he doesn't get bored," he'd remind everybody. "Let's keep up with our schedule. Get the guests, the singers, and the dancers out on stage *before* they're due. Keep it moving. I don't want Dean to have to wait for anybody!"

Dean was putting all his trust in Greg now. Their "handshake" contract was moving along handsomely.

I still felt shaky at those orchestra rehearsals every week — Dean's first awareness of the show and the moment of truth for all of us. Sometimes I couldn't stand the pressure and I'd wait down the hall to see if he was on his way. Maybe I could check his mood.

He'd drive up in his Dual Ghia, Mort Viner his only passenger. At the end of the day they'd reverse those positions. His star status had allowed him a special parking space just a few steps from the artist's entrance.

Close together, he and Mort would walk through the big doors, wave to the guards, and plow past the studio tours and the workmen from other NBC shows, always in a determined straight-ahead gait that seemed to tell passing admirers that these two guys really had to be on their way, that they might be late if they stopped to chat. It was a very studied, psychological, and successful way to get through the crowds.

I'd watch them carefully as they headed for Dean's dressing room just opposite the entrance to Studio 4. It was the only one Dean ever had at NBC. Certainly not the

Liberace, Dean, Carol Lawrence, Bob Newhart, and Dom DeLuise

studio's biggest or most elegantly furnished but a comfortable two-room suite with couches and chairs in the main room, a sleep sofa and lots of mirrors in the dressing area, plus a bath with shower. Dean would only spend a few hours there, his only requirements being a small bar with lots of ice, soda, Scotch, soft drinks, and a large color photograph of himself with Cardinal Spellman.

A few minutes before 1 p.m., like clockwork, Mort would come out and ask if it was time to step to the orchestra.

"Anytime," I'd assure him. "We're ready."

Les and his men knew the routine well. Promptly at one o'clock Dean would amble out, wave to the boys in the band, take time to tell them an R-rated joke, and then look over at me.

I got down to business fast. "Don't make Dean wait." The football game was more important. I'd weigh my words carefully, explain a medley or song, with whom he'd be singing it, and how it fit into the show. He'd listen carefully, nod his head as though he understood every word, and wait for me to give Les the sign to go ahead.

That little nod each week would tell me if I'd done my job. If the following half hour or so went smoothly there'd be a great sigh of relief. The guesswork for that week, at least, was over.

If my little chat with Dean was short, it was more than anyone else had with him. There was a sort of unspoken rule: Don't talk to Dean, don't bug him, let him breeze in and out without any interruptions. He in turn never gave anyone a chance to start up a conversation. Do the job, and get out.

Greg once asked associate producer Norm Hopps to come down from the control room and tell Dean about a small change in the script that wouldn't be on the cue cards.

"I'd be glad to," Norm answered, "if someone will just introduce us."

He ran down to Dean's dressing room

and announced, "Greg said we have to cut thirty seconds out of the next sketch." Dean stared at him and said, "I don't know who you are, pal, but if Greg says so, okay."

Those weekly orchestra rehearsals were carefully planned. I'd tell the stage managers to keep the guests away until we'd gone through Dean's two solos — his opening song and the ballad. That took care of six minutes.

Just as he was finishing the last bars, I'd wave to the stage manager to bring over, say, Peggy Lee. I'd allow enough time for them to kiss and exchange a few niceties (after all, this was their first meeting), and when it was diplomatically correct, I'd nudge Les to raise his baton to start their medley. Mustn't let Dean get bored.

The Dean Martin *image* was also carefully set up. A fresh cigarette from Jay Gerard, who attended Dean in that respect, a drink in his hand brought from the dressing room in case he needed to douse the tonsils, and a gold sweater to perhaps certify that he had actually just come in from the golf course at Hillcrest Country Club.

"I don't need this job," he'd tease. "I've got $500 in five different banks."

He and Peggy would watch the cue cards and go through their medley. I'd stand back

Me and the Andrews Sisters

to see if anything gave Dean a problem. If there was a sign of concern on his face or if he stopped momentarily, I was prepared.

"Oh, that should have been Peggy's line," I'd lie. "The cards are wrong. Sorry, Dean." Peggy was prepared during the week to take over if Dean didn't know a musical section. Les knew not to stop the band, to just keep playing.

"I could have phoned that one in." Dean pretended the whole thing was a snap.

"Okay, Peggy, thanks a lot. Bring over Paul Anka," I yelled. Same routine. "Keep Sheila MacRae out of sight. She's pushy. I'll get to her and Gordon MacRae last."

"Just one more number," I'd tell him and

ask Gordon and Sheila to come over. I knew that Dean liked him and tolerated her.

"Gordon is a great guy and the best singer around," Dean confided to me. "What's he doing with that untalented broad?" I'd signal Les to start the music. Sheila almost immediately questioned some of the band passages.

> *"The dame . . .*
> *Who is playing a game*
> *With a guy . . .*
> *Who is basically shy,*
> *While the star . . .*
> *Wanders off to the bar . . .*
> *That's entertainment!"*

I'd tell Les to plow through whatever song they were doing, promising Sheila we'd fix things later. They built to a big finish and Dean headed for the dressing room.

"What's the score?" he muttered to a stagehand as Sheila strutted like a linebacker directly toward Les and the unwary orchestra.

Back in the dressing room the scene must have been a show within a show. Harry Crane, our second writer, doubled from *The Andy Williams Show* and a privileged guest, was on hand to keep Dean "up," tell him stories, and, if need be, explain the cue cards.

Those cards were brought in by Barney McNulty, who was advised by Greg and Harry to move the cards and never speak to Dean. That wasn't easy to do, of course, and Barney couldn't resist throwing in a few suggestions to help Dean. As time went by, Barney became as important to Dean as Greg and I were. The original "Cue Card Guy," Barney began his long career as an usher (later called "pages") at CBS. One day on a live show, Ed Wynn was having a problem remembering his lines. He had made so many mistakes the producers were afraid they were going to have to throw out the routine. Barney, however, dashed out of the studio, ran across the street to a hardware store, bought some cardboard and a marker, and printed Ed's lines large enough for him to easily get through the sketch. He saved Ed and the show and gave birth to TV's famous cue cards. Thrilled with his success, he quit his job as an usher and started the cue card business.

There were no windows in Dean's dressing room, just a monitor specially hooked up to all the local TV stations and also a direct line to our stage. The latter was only turned on, however, when Greg wanted

Dean to watch something he'd be involved in later. To provide a good NFL contest for him to watch, NBC supplied feeds for almost any game Dean wanted to see that particular Sunday.

Our run-throughs were carefully scheduled so that Dean wouldn't have to make an appearance on stage until halftime. Greg would even assign someone to pay attention to Dean's game monitor in order to avoid any disturbance. Later, if there was something important for him to see and the game was at a standstill, Greg would ask the control room to switch to the stage monitor so that Dean could watch me go through a sketch or a song in his place. I suppose there must have been a few laughs as well over the way I was playing Dean.

"Okay, I got it," Dean would report. Magically the monitor would switch back to "third down, a yard to go."

After the first year, we dispensed with dress rehearsal and went right to the taping after camera blocking. "Don't let him lose interest" was the common cry. It was true that the first time Dean did anything was usually the best, most refreshing, and most fun. We decided to share that with the audience.

It was important that Dean should be in

the best possible mood for the show, even if his favorite team lost. So Greg would bring him out to meet the audience three or four minutes before the show began. Dean would do several stories from his nightclub act to warm up the audience and himself. When we reached a comfortable laugh peak, Greg would pull him over to his first position.

"Point the Italian where you want him," Dean would say, showing his total confidence in Greg. Dean would look for the cue cards, read, goof, make the most of it, and it was all, well, our weekly pattern.

Cleveland Amory put it this way in a midseason review in *TV Guide*:

Whether Mr. Martin does his show the way he does because he thinks it's better that way or because he's lazy — well, it doesn't matter. The fact is, it is better that way. Indeed, it is high time to say that Mr. Martin now has the best variety show on the air. In the first place, it offers genuine variety. Since it seems to be the most difficult element to get in a variety show, there is actually variety in the variety. In other words, there is pace. And the pace on this show is not just good, it's darn near perfect. A formidable mixer on any level, Mr. Martin has a really remark-

able ability not only to mix his guests but also to mix with them — and often make them look better than they did alone. He's able to bring it off with such self-conscious unself-consciousness or such unself-conscious self-consciousness — and again, it doesn't really make any difference which.

Dean and Gordon MacRae

Before the season started, research departments in advertising agencies used charts, graphs, slide rules, and computers to assign the probable popularity of each of the 97 shows coming up in prime time. Ranked 97th — in other words, the most likely loser and least likely winner — was *The Dean Martin Show*. The idea of Dean lasting out the season is most improbable, they said.

Newsweek was quick to report how wrong they were:

Want success as a TV variety show star? Relax. Perry Como and Andy Williams proved the rule, though their effortlessness had to be rehearsed. To produce a 60-minute program, Como tapes for as much as two days, Williams for up to eleven hours. Now both of them are surprisingly being out-relaxed by a competitor so genuinely casual that he can't be bothered with a real re-hearsal or even a retake. As a result, The Dean Martin Show is hoisting itself ever higher in the ratings and is the closest thing on the air to the free and easy spontaneity of old fashioned live television.

It was difficult for some guests to work this way. Even though they were well aware of the impact of Dean's style and wanted to be part of it, they were scared to death of what would happen when they got their one and only chance with him on our Sunday night tapings. They had to be soothed and pampered and convinced that it would all work out fine. And when it did, they were amazed.

Dean, of course, was completely unaware of their insecurity about working with him. As a matter of fact, he was unaware of anything we were up to before he got to the studio.

"Just show up," Greg told him. "We'll do the rest."

So all week long we rehearsed. The guests and singers and dancers had to be constantly reminded that the show was only casual-*looking*, that Dean made it that way. The rest of us had to know exactly what we were doing and be prepared for any surprise. And oh boy, Dean could surprise us!

Dean loved it that way. He happily did what he was told, and seldom balked at anything thrown at him. No temper, no interference, he accepted whatever the show asked him to do — but he always did it his own special way. All he had to do, really, was be Dean Martin once a week.

Although there were no official orders to the effect, it was understood that Dean's dressing room on Sunday was off-limits. We all knew by now that bugging him with anything concerning the show would take away from the fun he'd have finding out for himself.

Occasionally one of the guests would knock on his door just to be sociable.

"Dean's sleeping," Jay Gerard would report.

"Sleeping? My God, he goes on the air in an hour."

"I know — he's resting up for it."

The bit about Dean's always wearing a tuxedo on the show, no matter what the rest of the cast wore, was not a sign of laziness on his part. And it wasn't a Cary Grant sophistication with him either. Tuxedos seemed to be created for Dean. His tailor, Joe Kleinbarth, was a genius and knew what suited him best. Dean looked more comfortable in a tux than most people do in their pajamas.

And his barber deserves credit as well. Dean's hair was always just right. In total, he was the best-dressed, best-looking man in town. It was easy for the rest of us to further that image when Dean carried it off with such style and class himself.

Sometimes Greg would call him to the stage and for some reason we might not be quite ready for him. They'd have to work on a camera or maybe a prop wasn't set. Greg and Dean would instinctively go into a little routine.

"Sorry, Dean, a short delay," Greg's voice would boom down from the control room.

"It's okay, buddy. Whenever you say."

"You hungry, Dean?"

"Yeah, I could eat something."

"How about a sandwich?"

"Sounds good. What kind have you got up there?"

"There's a tongue sandwich left."

"Tongue? No thanks, I don't eat anything that comes from an animal's mouth."

A pause to let that sink in.

"How about a couple of eggs?"

The audience lapped it up. It was just off-color enough to make them all giggle with delight. Something naughty they could tell the folks back home.

"Dean, your wife just called. She says when you get through here to get your tail down at the Biltmore Hotel."

"Tell her that's where I always get it."

Now everybody was in a good mood for the show.

In spite of the dangerously few hours in our Sunday schedule, we always seemed to wrap up early. And Dean was the first one out. When we finished with him onstage,

86

Mort Viner would take him right out the back door of Studio 4 still dressed in his tuxedo. Dean was in his car and on his way home while we were playing the closing theme music.

Since executive producer Hal Kemp constantly overbooked the show, we were forced to make the best of some awkward situations involving people we knew were not exactly right for Greg's master plan. Okay for some of the other shows, maybe, but not ours — we were doing something different.

Greg never wanted me to go to any of the guests. They were to come to my office. "Keep them in their place," he'd say. "Never forget that you're in charge now."

Dean and the chorus girls during a station break

But Allan Sherman had called to say he'd be unable to come over and would somebody please go out to his house?

Sherman was a writer who had been measurably successful with his parodies of popular and classical material. Because of his big hit "Hello Muddah, Hello Fadduh," he was popping up on TV variety shows with reasonably humorous special lyrics and a questionable personality to carry them off. He was termed an "unusual" talent.

"We're going to have to coddle him a little," Greg told me, regretfully. "He's an old friend of Kemp's." So off I went to his big house in Beverly Hills.

It was an exceptionally hot day and when I arrived and announced who I was, the maid escorted me to a chair in the hallway to wait. I could hear voices of what sounded like a large noisy family at lunch. "Damn, I could be doing a thousand other things," I thought as I waited. And waited. And waited . . .

I had called personally from the office to verify my arrival time, so I felt there was no possible excuse for this uncomfortable delay. After an hour went by I decided to get up and have a look around. I just wanted to find somebody, anybody.

I followed the noise to the dining room,

where a zoo of unruly kids and noisy elders were clawing and grabbing at a huge table full of sandwiches and fixings.

I barged in. "Excuse me, I've been waiting for Mr. Sherman for almost an hour," I said firmly. Nobody paid the slightest attention to me. Lunch was more important.

Finally a small boy near me went to the dining room door. "DADDY!" he screamed up the stairs, then quickly returned to the food. "Good boy," I thought, "roll over . . . sit . . . play dead!"

There was no response. Just when I decided this was all hopeless, someone came over and asked, "Can't you just wait in the hall?"

"I've done that," I shot back, "for quite some time."

"Then why don't you look upstairs for him?"

Ah, some progress. I hiked up the large circular staircase, saw a door open at the top, and knocked.

"Yeah, who is it?"

I announced myself and walked in.

"Where the hell have you been?" he asked gruffly.

"Okay, calm," I thought. "A favor for Mr. Kemp. He'll owe me one after this."

Sherman was lying on his bed wearing

only a pair of shorts, sweating profusely with the heat, a phone in one hand, pad and pencil in the other, and several cans of ice-cold beer close by.

"I bet he doesn't offer me one of those," I said to myself. He didn't.

He put down the phone. "I got the script. I don't like it. Those writers are lousy. This stuff is crap. I'll let you know what parodies I'm doing. I've got some calls to make. The maid will see you out."

I was tempted to say, "Listen, you big fat whale, I came to discuss what you and Dean are going to do on the show, not how busy you are." But what was the point? I remembered what Greg had said about this being a favor and that he wasn't going to do much with Sherman at all.

"But maybe we could talk for just a moment about . . ."

"This ain't the only show I'm doing, buddy. Do you mind?"

He waved me toward the door.

I managed a "you son of a bitch" under my breath, which I'm sure he didn't hear over his new phone conversation. Maddening as it was, it was a relief to get away from this pompous *artiste*.

I reported back to Greg, who wasn't the least bit surprised and actually found the story rather funny. So did I after awhile, especially when I heard that Greg wasn't even going to roll tape when Allan was on the stage.

"And that's the last time I send you to anybody's house!" said Greg.

A few weeks later he went back on his word. This time it was Lucille Ball who needed special treatment. She was busy taping her own show and humbly asked if I'd mind meeting her at her studio. We all knew what a perfectionist she was and we were concerned about what sort of reaction she'd have to our easy-going, freewheeling style. She could be difficult, too, if everything wasn't just so. And how would she take the fact that there'd be no rehearsing with her costar?

"It's all up to you," Greg said. "Just explain how we do things around here. She ain't gonna bite you." Greg liked to use bad grammar to make a point. "I know her. She'll understand." Somehow that "bite you" phrase lingered in the back of my mind.

It happened that Kate Smith was booked for the same show. Kevin and I chuckled about what an unlikely trio that was going to be, Lucy, Kate, and Dean. "We'll find something easy for Dean, something the ladies can do all the work in."

We came up with an old-fashioned vaude-ville medley in which Dean and Kate would be the production singers and Lucy the klutzy chorus girl who messes everything up. It seemed perfect. Now we'd just have to convince Lucy that it was perfect.

We arrived at her office at Desilu Studios right on time. Greg had scheduled this meeting a week ahead of schedule to give us enough leeway to change plans if this little gem went up in smoke. We told Lucy our idea.

"So? That's the kind of thing I do every week on my show."

"But not with Kate Smith and Dean Martin," I threw in.

She raised one of her red eyebrows. "Okay, what are the songs?"

Hey, we got that far!

" 'Bird in a Gilded Cage,' 'Shade of the Old Apple Tree,' 'Wait Till the Sun Shines, Nellie' — stuff like that."

"And what am I going to do with them?"

"You're out of step with the other chorus girls, you just happen to knock down the scenery behind Dean and Kate, you're a real klutz. You bite off more than you can chew."

"There's gotta be something more than that." She immediately came up with ten more ideas, some of which we had considered earlier but didn't want to go too far with her. Fortunately, she did.

There was one bit we concocted during "Asleep in the Deep" where she was to ring a huge church bell after Kate sang the line "loudly the bell in the old tower rings." The audience would only see the huge rope Lucy would yank. After a pull, she'd disappear up into the rafters. During rehearsal she insisted that Kevin try it first. He did, the rope gave way, and he fell to the floor. Although he wasn't hurt, we thought it would surely scare off Lucy. Yet after repairs and some testing, she insisted on going up and down *more* than one time. Once wouldn't be funny, she thought, but three times would.

There was also a musical passage in "I Don't Care" that Lucy had difficulty with all week long, right down to the orchestra rehearsal. Each time she got to it, Kate would try to help by booming out the correct notes in her ear. Over and over Kate would bellow the line, much to the famous redhead's annoyance.

"Get that broad out of here!" Lucy whispered in my ear.

I told Kate not to be concerned about it, that we'd work it out later. But she wouldn't let it be.

Kate Smith and Dean, with Lucille Ball being yanked off her feet

"It's so easy, Lucy, don't you see? Here, let me show you again." Kate would go over the section perfectly every time. I thought I caught a glimpse of Lucy eyeing Kate's jugular. That was it, I thought, and the headlines flashed before me: "Big Red Bites Big Belter."

We almost had to change songs. Lucy got more and more furious with Kate, then stopped trying entirely, insisting we go on with something else. The next day I was ready for some lengthy work with her, but she didn't want to discuss it and kept saying she'd get it. She did get through it, shakily, when we finally put it on tape with Dean.

We all thought it was fine, but Kate kept pulling at Greg and me saying, "Aren't we going to do it again? Aren't we going to do it again? Lucy missed a couple of notes, you know." Lucy didn't hear that, fortunately, and Greg walked Kate over to the side to plead with her to let the whole thing drop. Dean, of course, was already in his car and on his way home.

Coming up the following day was a lady we all knew from the days before her huge success on *The Dick Van Dyke Show*, Rose Marie. She hadn't used her singing voice lately, and we wanted to show America that she was one of our better belters. She had brought over a couple of real "up" tunes, insisting that was her forte, but Greg and I had other plans.

We diplomatically convinced her that although she was not a great beauty, there was some value in making use of that. Surrounded by lovely ladies and handsome men at a dance, she'd be the wallflower everyone ignored.

"You'll have to act it," Greg kept saying. "Anybody can sing a rousing song and get a hand. Why don't we go for a little tear in the eye?"

She thought it was a terrible idea, that she couldn't possibly bring it off, and kept asking herself what she'd gotten herself into on this strange show. Greg kept pushing and after some hefty ego inflation talked her into it. "If I can still do 'I Got Rhythm'," she added.

To the strains of "Little Girl Blue," she tearfully allowed man after man to pass her by for prettier faces, singing Rodgers and Hart's soulful song amid the rest of the throng. But each time we rehearsed it she broke down and asked if she could talk to Greg. "This isn't going to work," she kept repeating. "I can't tell you why, but I just can't do it." Maybe we'd pushed her too far, or embarrassed her.

Greg made regular visits to the rehearsal hall and repeatedly talked her into the number.

Taping came and she performed the hell out of it, wrenching tears from everybody in the house. We had devised a way for Dean to find her, sing a chorus of "Smile," and cheer her up. She grabbed his handkerchief to dry her eyes. She was totally drained as Dean held her in his arms and we simply forgot about "I Got Rhythm."

96

Several weeks later she sent me a note that explained a lot:

Dear Lee:

I know it's taken a while to write this, but I really was so emotionally shaken up after doing your show, it's taken me some time to 'come down.' However, I want to thank you for your kindness and help . . . and 'putting up' with me. Most of it was a personal thing for me, the death of my husband just before you saw me . . . quite a long story.

Always, with love . . .
Rose Marie

Oh Rosie, I wish I had known. You are the epitome of the word *trouper*, and I love you.

The Krofft Puppets had been a weekly feature on the show; something that seemed to have come with the original contract. As popular and unusual as they were, it became more and more difficult to find things for them to do with Dean. Nothing seemed to work. Most of what we put on tape never made it to the edited show. The Kroffts were finally let go before the end of the season.

Sid Krofft felt it was because the heavy mail they'd been getting was a threat to Dean. In actuality their departure was

Dean and
Lainie Kazan

simply due to the fact that Dean felt uncomfortable working with "dummies," as he referred to the puppets. It showed. Dean's reactions were forced. So Greg wished them well and bid them goodbye.

We'd have to find other departments for Dean. If not girl puppets, how about *live* girls? After the Kroffts left we added four more girls to the chorus and began a series of musical station breaks, always with the attitude that Dean was the All-American super ladies man.

His drinking act had come from his nightclub days. Now we were adding girls. We began building a new image, a new look. That heralded the birth of "Booze and Broads," synonymous with Dean Martin.

What to do with a big star like Bob Hope?

Yes, he'd agreed to come in for rehearsals, and yes, he'd be doing his own monologue. But he and Dean together would need some special handling.

Some clever repartee from the writers would be obvious. But what musically? Juliet Prowse was on the same show and had proved earlier that she could handle Dean nicely. So Juliet to the rescue! We used a French and English routine ("French Lesson" and "Brush Up Your Shakespeare") with Juliet teaching her two Romeos more than they'd perhaps want to know.

Bob was understanding about having to rehearse all this without Dean, but when we taped it Bob goofed and Dean didn't.

"Someday I'm gonna get my own weekly show," Bob ad-libbed, "so I can do whatever I want."

"What do you mean?" Dean added quickly. "Any show you're on *is* your show."

"But how come they're laughing at your straight lines?"

Dean was gracious with all his guests, but he allowed Bob plenty of elbow room. They genuinely liked each other.

To Bob and other comedians, Dean was the perfect straight man. To the ladies he was the perfect gentleman. He made everyone feel at ease. But there was one silly rule they had to follow. Although it was never mentioned by Dean, we discovered that he was more comfortable standing on a performer's *right* side, probably because of all those years with Jerry. Our staging always kept him on that side.

Barbra Streisand was born in Brooklyn. So was Lainie Kazan. Barbra went to Erasmus Hall High School in Flatbush. Lainie did, too. Barbra was a big star on Broadway in *Funny Girl*. Lainie was a twinkling little star whose first good job anywhere was as Barbra's understudy in *Funny Girl*. But they rarely talked. Barbra knew Lainie was good, but Barbra never missed a performance, except for one Saturday when she called in with a bad case of laryngitis. So Lainie went to work. She phoned everyone she could think of who might be important to come see her at both the matinee and evening performances, the only time Barbra was out during the whole *Funny Girl* run.

It paid off. The right people came and loved Lainie. Word got around about her and it changed her career. She was brilliant in the Fanny Brice role and the public discovered that she had an incredible voice, was prettier than Barbra, and very sexy.

Greg somehow heard about this and immediately booked Lainie on our show. Coming in with only her one-shot reputation on Broadway, none of us were quite sure what to expect. What we got was a superb song stylist and a new playmate for Dean. He adored her instantly. Their little moments together just passed the NBC censors. I knew they'd have fun with slow, sexy versions of "Cuddle Up a Little Closer" and "Put Your Arms Around Me, Honey, Hold Me Tight."

Lainie gave *The Dean Martin Show* full credit for establishing her with the public outside Broadway. She came back many times, always easily maneuvering from high-powered versions of things like "I Loves You, Porgy" to a totally delightful talk-and-song session with Dean. They dated several times, but Lainie insists he never asked her to hit the sack. "We just had a good time together," she reminisced. "He was always the gentleman, never laid a hand on me. Greg, on the other hand. . . ."

Two years after Lainie's debut with us, she was booked on the same show with one of her idols, Ethel Merman.

So I put together a showstopper medley for Ethel and Lainie and Dean, giving everybody a chance to belt out some big standards. Dean remained steadfast in the middle of the two belters, giving most of his attention to (and getting help with the songs from) Lainie. On their last big note, he gave Ethel a polite peck on the cheek and grabbed Lainie in a meaningful bear hug. The audience understood. So, I think, did Ethel.

Our final show of the first season must have been a Hal Kemp afterthought. Unlike previous shows, this one had only four guests: Bill Cosby, Dorothy Louden, Guy Marks, and Liberace.

Dorothy had come from nightclubs and nobody knew who she was. She was probably a "cheap act" for Mr. Kemp to book in 1966. But what a luxury to have only four stars (and a "paralyzing pirouette" by Tanya the Elephant) to take care of!

I had worked with Dorothy in New York, where she did a hysterically titillating flirtation medley. There was a line of chorus boys she would try to vamp and coerce with her fluttering eye, puckering lips and undulating body techniques. Yet one gentleman could not be swayed by any of these feminine charms. An obvious position for Dean to be in. He wouldn't have to do anything but sit there.

But Dorothy questioned whether Dean

could carry it off, what with no rehearsals. Well then, maybe Bill Cosby could do it? No, there was still that black-and-white problem on TV in the '60s. A white lady seducing an African-American man? Too risky. How about Liberace? Another problem. Would he be *too* convincing as a man who ignored her advances? Guy Marks? Not available for rehearsals and out of town until taping day.

We had hired extra men for the number, twelve in all, thanks to money saved from a small guest list. We tried each one, but no one could do it with a straight face. Then Greg insisted I do it.

So there *I* sat on a bar stool along with eleven other tuxedoed beaus on their stools as Dorothy slalomed up and down our row.

> *"Every little movement*
> *Has a meaning all its own . . ."*

She batted huge false eyelashes and two guys cracked up. She contorted her lips like "The Blob" in heat and four more broke up laughing. One by one they fell to her vivacious seductions. I was next. She attacked again and again; I ground my teeth. As she mercilessly wriggled and squirmed, I gripped the stool and tightened my stomach.

Ethel Merman, Dean, and Lainie Kazan

She mussed my hair, blew in my ear, and undid my tie but I wouldn't laugh. I nearly wet my pants, but I didn't laugh through the entire twelve-minute routine.

Dean's comment? "Why didn't you let me do that?"

Actually he would have been very good at being totally deadpan. But this was still our first year, and we didn't think he was quite ready for that sort of thing — yet. Instead he'd have to be content with an end-of-the-season medley, another twelve minutes, with Liberace. Lee (as his friends called him) had turned out to be an especially good partner for Dean, agreeing to do anything, including replacing Ken Lane and accompanying Dean for his parodies and ballad. Lee also took in stride Dean's constant comments about his sequin-heavy wardrobe.

"I had this jacket made especially for your show."

"I didn't figure you got it off a rack."

Mr. Rhinestone and Macho Man. They both played it to the hilt.

Greg hated special material. "The Midwest won't understand it" was his argument. So the next best thing was writing special lyrics to familiar tunes. Dean didn't have time to learn special material, of course, but once he was in a song he knew, he'd carry on as long as the cue cards were there, old lyrics or new. Since this was our last show of the year, it seemed appropriate to make some sort of acknowledgment of that. Liberace's musicianship held together the twelve-minute medley, which used a lot of name-dropping to the tune of "Pretty Baby" to avoid special material and summed up Dean's first successful year on NBC:

Liberace: *Eddie Fisher, Carol Lawrence, Patty, Maxine, and La Verne, Miss Mahalia*
Dean: *Gordon and Shay-lia*

Liberace: *Louis Prima, Polly Bergen, Patti Page and Richard Hearne, Hope and Lucy*

Dean: *Allen and Roosi!*

The medley ended with special lyrics to "Makin' Whoopee":

Liberace: *You came around and did your show*

Dean: *And I was bound to make it go*

Liberace: *You did it weekly*

Dean: *And so uniquely*

Dean and Liberace: *With no rehearsals!*

Liberace: *You never worried those many weeks*

Dean: *Well, when I'm hurried my body creaks*

Liberace: *He gets so tired*

Dean: *I'm not required*

Dean and Liberace: *To make rehearsals!*

Liberace: *He gets upset at no time*
He's not a nervous chap

Dean: *And just before each show time*
I always take a nip — er, nap!

Liberace: *He sings his song and then he speaks*

Dean: *I got along for thirty weeks*
Here's what you do, pard . . .
Just watch the cue card.

Dean and Liberace: *Who needs rehearsals?*

Well, *somebody* did, but not Dean. Gradually we were able to work him into more and more spots on the show, the ratings climbed higher and higher, new contracts with NBC were in the works, and Dean actually spent less time at the studio than he did before. Gone were the weekly meetings at his home and dress rehearsals. We simply taped the first time he put on his tuxedo, mistakes and all. Dean had a great time and it was infectious. He trusted Greg to make all the decisions on redoing anything that went terribly wrong, but usually Greg would somehow leave it in.

It was the *style* of the show that viewers loved. Every Friday morning people all over the country would ask each other, "Did you see what Dean did last night?"

To celebrate our success Greg decided there should be a lavish party on Stage 1. Mr. Kemp agreed that his budget could stand it, especially since he was about to sign a lucrative new contract with the network.

But would Dean show up? We all knew how distant he was. It seemed unlikely that he'd stay after the taping to socialize. But there he was, charming everyone. Cast and crew walked up and introduced themselves for the first time and Dean even managed to make conversation with excited relatives of all the "behind-the-scenes" staff.

That night we all got to know him just a little bit better. Well, at least we found out he was real. He had to spread it around, what with 150 studio workers on hand, but he proved to be quick-witted, warm, and outgoing. He seemed to realize he owed them all something, his presence at least, and he stayed for a good two hours before his ever-attentive manager Mort Viner sneaked him out the back door.

That night was something we all had to savor. It was the last party he ever attended at the studio.

Still, there was plenty to celebrate. It was like graduation day, except we knew we were all coming back.

3 The Second Year — (1966–1967)

Up, Up, and Away!

The Dean Martin Show was now firmly established as an out-and-out ratings hit. NBC was so happy, they signed Dean and Greg for three more seasons.

Greg was so pleased with my work that he offered me any credit I wanted. I chose the "Special Material By" credit because that was what Earl Brown, Billy Barnes, Artie Malvin, and Ray Charles used on their respective variety shows in similar jobs.

As that first year went by, I advanced from writing a couple of medleys for Dean and his guests to putting together production numbers for the guest stars and, incredibly, standing in for Dean in every rehearsal right up to taping.

In those days every hit show had a "summer replacement," usually a similar type of program that kept the time period occupied with the same audience until the season started again in September. Dean had taken a firm grip on Thursday nights at ten o'clock and his summer replacements were always preceded with *Dean Martin Presents* . . . to assure everybody that he'd be back in the fall with the booze and the broads.

Summer shows meant low budgets, usually a third of what the regular season shows cost, the network arguing that only a third of the viewers were watching during the summer months. Commercial prices went down, too. Consequently, there was always a search for new faces, talented people who would be happy to work for next to nothing in order to showcase themselves on network television. For the creative staff it simply meant less money and harder work, trying to make the shows look and sound as good as their fall counterparts, in spite of production trappings down to the lowest monetary level possible.

Dan Rowan and Dick Martin had made an appearance on Dean's show, among others, and were constantly being referred to as the new Abbott and Costello, Olson and Johnson, or even Martin and Lewis. Their comedy was more cerebral, never raucous. Dan and Dick had ideas of their own, too, about doing a variety show — short,

Rowan & Martin

topical sketches with a company of regulars.
A familiar ring? What we all came up with
was a sort of prelude to *Laugh-In*, but with
Greg's added touch of having the sets
placed and replaced by the cast on camera,
a nice flow. And we needed some more new
people.

"What about DeLuise?" Greg asked.
"What kind of guy is he? Could he be a part
of this with Dan and Dick?"

There was no question in my mind. Dom
and I had worked together on Carol Burnett's 1964 CBS series, *The Entertainers*,
and it seemed there was nothing Dom
couldn't do and be hilarious doing it. "He'd
be terrific," I said.

Still Greg insisted on an interview, even having Dom audition some of his material. Dom flew out from New York shouting, "I don't mind! I don't mind!" — an exclamation that even applied to his low salary. "I'll do anything to be on your show — get you coffee, drive you to work, suck your toes." Greg dug Dom from that first day and knew his exuberant style would complement Dan and Dick.

No salary problems arose with the featured singers, either: Lainie Kazan, Frankie Randall, and Judi Rolin, all of whom were as eager as Dom to do the show. There'd be a chorus of young singers and dancers who would also be available for Greg's scene-shifting ideas.

I called one of my singers from *The Entertainers*, Melissa Stafford, to come out from New York and be part of the chorus. She was a beauty, a Marilyn Monroe look-alike with an endearing naiveté. On first meeting her, most people were sure she was totally unreal. But not Greg. He found her so refreshing that he opened the show with her every week — close-up to full figure as she brought out the cast and the movable scenery. From that moment on, Greg never did a *Dean Martin Show* without Melissa. She and soprano Loulie Jean Norman were his good luck charms.

After our opening summer show the critics were ecstatic. Robert Sylvester wrote in the *New York Daily News*:

Dean Martin has the best TV show in my book and I'm delighted to report that also in my book his summer replacement show is a smash. The stars are Dan Rowan and Dick Martin, who have been around for years without getting very far. For their first show they had some hilarious sketches and fine production ideas. The big thing about this show — outside of the quality of its parts — is the finesse in which the parts are woven together. Here's one show which ought to make it during the winter, too.

The ratings were sky-high, too. Under the headline "Dean Martin Summer Show Tops All Opposition" — the *Hollywood Reporter* announced that the show had been No. 1 for the summer:

All the others are in flopsville position. To tabulate a few: 54th position to John Davidson's Kraft Music Hall, *68th for* Continental Showcase, *59th for* Mickey Finn's, *88th for* The Face Is Familiar, *and very few viewers are taking an affectionate look at the* Wayne and Shuster Take An Affectionate Look *series because this one's at the bottom of the numbers.*

All the reviews mentioned Dick and Dan and Dom, but also had nice things to say about the rest of the show, particularly the finale, in which Greg insisted I use the entire cast.

Those finales were a bit of nostalgia close to the hearts of all of us, something we all wanted to get out of our systems. Dick, Dan, Dom, Greg and I were all brought up in the Big Band Era, with Les Brown being one of them, so each week we lovingly recreated

Greg and me standing in for Phil Harris and Dean

114

Dean and
Buddy Hackett

the sounds of Glenn Miller, Tommy Dorsey, Artie Shaw, etc. What fun to dig out the old 78s and run them down! Our pianist Geoff Clarkson and I would sketch out a song for everyone in the cast, then turn it (and the sometimes quite scratchy recordings) over to arranger J. Hill, who painstakingly recreated the original sounds.

During the time between Dean's last regular season show and the start-up of his summer replacement, we lost Kevin Carlisle to Broadway. His contributions had been terrific that first year and it wasn't easy to see him go. But he was determined that his next step was back to the "Great White Way," so Greg wished him well and called in an old friend, Bob Sidney, to take over the choreography chores.

"He staged some numbers for Dean and Jerry in their Paramount movies," Greg explained. "Maybe Dean will remember him."

"You're either going to love him or hate him," Greg warned us all in advance. "But to me, he's the best stager of musical numbers in the business. He always makes the stars look better than they would otherwise."

I found out pretty fast that Sidney, as everyone referred to him, could be either stingingly evil (usually upon first meeting) or understandingly sensitive, sometimes both at the same time. His admirers insisted that his put-downs were simply his brand of humor, and once people figured that out they became fans, too. And I soon discovered what Greg meant about Sidney's work with stars and nondancers. He gave full attention to the personality out front, never covered them up with fancy steps from the more experienced hoofers around them, never allowed those dancers to kick higher than the star, never staged it so the star was danced off in the middle of the number so his dancers could show off for sixteen bars. When he joined *The Dean Martin Show*, Sidney was particularly sensitive to Dean, making sure everything was carefully staged so that Dean could merely walk into the middle of a routine and not have to think too much about what he was supposed to do there.

The guests and the dancers were directed to push Dean this way or that, and Dean would instinctively go in the direction pushed until stopped by someone else. Standing in for him every week, I made sure I didn't do too much on my own, waiting for somebody to push or shove me as they would have to do with Dean. The right push was sometimes critical. We really only had one chance with Dean — the taping!

Sidney and I worked well together. Whoever got an idea first, we'd just build a number as we went along, never failing to change it until the last minute if we thought we could make it better. Chattering away, he'd make suggestions for additions and deletions all week long, even after rehearsals were over. We'd leave NBC separately and on my way home I'd run into him at the grocery store.

"And another thing . . . ," he'd say, picking up the conversation where we left off at the studio. Then he'd invite himself to dinner.

As our second season began, Sidney wanted to make sure, since we were not a "girlie" show, that whatever the ladies did

Me standing in for Guy Marks with Peggy Lee, Dean, Joey Heatherton, and Sid Caesar

around Dean had some semblance of taste. "They're not going to paw him and pat him like he was a cocker spaniel," asserted Sidney. "It's much better to do things by innuendo." He created little bits week after week that had taste, humor, *and* sex, keeping the girls just out of Dean's reach.

And Dean knew how to move with his costars, to make them look good no matter what he did or didn't do right. He was instinctively a brilliant showman. And nowhere else was it more apparent than in his dancing moments. Even when he made a mistake there was a certain grace about it. So much so that, with the tuxedo to help, he always looked totally at ease.

★ ★ ★

When I arrived at the office to talk about Dean's second year on NBC, Greg motioned me to come down to his office immediately. Some sort of crisis, I thought.

"I'm doubling your salary!" announced Greg.

"My God, thanks."

"What are you thanking me for? Thank Dean. He needs you now, you know."

Greg was back to the phone and I withdrew with a smile, the meeting lasting about fifteen seconds. Greg was usually embarrassed about doing nice things for people. He embarrassed himself each year with my salary, which always went up. And I was sure Dean had nothing to do with it and knew nothing about it.

The year began with fewer guests. Apparently Greg now had some influence on Mr. Kemp, who would have preferred filling the season with jugglers and animal acts. Greg also seemed to have a voice in guest selection. It was clear that he wanted personalities who would work well with Dean and he felt he knew just who those people were.

The opening show was typical. Peggy Lee, Buddy Hackett, Rowan & Martin, Dorothy Provine, and Guy Marks were the guests, all of whom had played with Dean before and the good vibrations would be obvious.

The list suggested Damon Runyon to me, so I put together a rather long medley from *Guys and Dolls*, knowing that Dean had learned some of the songs when he did an album for Reprise with Sinatra and Crosby. Buddy was the only problem. He wasn't known for his musical abilities, so I gave him the talk-sing sections, leaving the real vocal work for everyone else. Without saying anything about this to the rest of the cast, they all supported him and knew this was a struggle, albeit a humorous one. They would hum the right notes in his ear when necessary, or sing along with him when all else failed. The result was a hilarious ten minutes with Frank Loesser's Broadway show, in what turned out to be perfect casting. Because of the publishing restrictions, though, we had to do it all in tuxes and long dresses. Anything resembling the original look of the show was forbidden. It didn't matter. This gang could have made it work in bathing suits.

Tender and vulnerable, Peggy Lee was treated by cast and crew with the kindest regard. She never asked for special treatment, but she got it because we all loved her. Her breathing problems were known to us all. An oxygen tank was a regular piece of furniture in her dressing room. I tried to limit her rehearsals, yet she seemed to regret that, wanting to be a part of what was going on at all times. There was the usual big song solo, of course, and the duet with Dean, who obviously had great affection for her as she did for him. It showed on camera.

We could always tell when Dean really liked a guest. He'd depart considerably from the cue cards and just say what he felt. At the end of their medley together, he held off the applause long enough to look into the camera and say, "A wonderful, wonderful person . . . the best in anything she says or sings . . . a true lady in every sense of the word. I love her."

Peggy's appearance on the show seemed to be a logical time to depart from Greg's insistence that all guests should come to me instead of vice versa. Before our first studio rehearsal, I called Peggy and told her I could drop by and go over her assignments with her. She was thrilled, so I hurried out to her Benedict Canyon home.

She welcomed me at the door in a gorgeous white caftan and thanked me for coming all the way out to her place but "isn't it much nicer this way?"

I was glad I came. Our appointment was scheduled for 3–5 p.m., but I didn't leave

Dean and Peggy Lee

until after 11:00. We settled on her solo songs and the patter with Dean. I went through her medley and the finale and she loved it all. She then proceeded to play some of her favorite jazz records for me while somehow managing to bring out a plentiful dinner. Her favorite wine, too, of course. As I was leaving she began apologizing for talking so much, a direct result, she said, of her attempts to quit smoking. The next day I sent her a pacifier from Cartier's. It was a small extravagance, but with my new salary increase I felt I could afford it and besides, Peggy was worth it!

The only problem I ever had with Peggy was really not her problem. Her vocal range was not compatible with Dean's. When it was comfortable for her it wasn't for Dean, and so forth. So I had to make key changes here and there so both would be happy, even though the end result may have sounded a bit disjointed to the public ear. But it didn't matter. The point was that they were enjoying themselves immensely.

Liberace, Bob Newhart, Dom DeLuise, and Carol Lawrence all made return appearances on our second show that year, and by now each of them knew what to expect from Dean. Bob took another of his record monologues out of the closet and revamped it to work as a sketch with Dean. Dom was all over the show, proving that two Italians can be crazy and funny without acknowledging the presence of cue cards. Greg liked having Liberace around, but wasn't thrilled with his insistence on playing still another of his glitzy pop classical piano pieces.

"I've seen him prancing around the stage during his personal appearances, titillating the old ladies with his dancing," said Greg. "Let's get him to do that."

It was true that Liberace liked to kick up

Dean, George Burns, and our choir

his heels now and then, so Sidney and I put together a satire of the old Warner Brothers musicals, a *42nd Street*-type production number with the girls. To our amazement, he loved it and learned several routines beyond the time step that made it look as though he knew what he was doing. Gene Kelly needn't have worried, but it impressed the audience *and* Greg, who forever reminded us that it was literally our *duty* to get our guest stars to do something they wouldn't ordinarily do on a variety show.

Happy as he was at the studio, Liberace would later confess that he was very unhappy with his private life, that he didn't have any "real friends." At the studio he was the friendliest of men. Crews loved him. I never heard a negative word about him.

"It's all a masquerade," he'd say sadly. "That's not really me."

Insensitively, I brought him back to reality with, "Speaking of masquerades, the decorators on Robertson Boulevard are having one next week. Wanna go?" Who would be more fun at a masked ball than Liberace in one of his fantastic outfits? He not only agreed, but immediately began planning what he would wear.

At the last minute he decided to go as himself instead of the Mad Hatter as originally planned. And of course the inevitable happened. "What a great outfit," everybody told him. "You're a dead ringer for Liberace! You look just like him!" Nobody believed he was the real thing.

He was extremely generous with gifts and dinner invitations and liked to invite Sidney and me to his palatial home in Palm Springs. "This place looks like a piano whorehouse," Sidney would say as soon as he walked in the front door. "There's one in every room. Why don't you put an ad in the *Free Press* for people with piano fetishes? I feel like I'm being attacked!"

★ ★ ★

Like Dean, Phil Harris made a career out of pretending to be the town drunk. Phil, perhaps, took it a little more seriously than Dean, but they both had perfected that image — the glorification of booze. Two new writers on the show, Rich Eustis and Al Rogers, had an idea for the two of them: they could engage in a tea party with all the English frills, but the audience would know that there was something stronger than tea in those cups. Girls in short maid outfits would naturally stroll in and out serving.

"One lump or two?"

"Dean, you dog, you know I'm driving."

Eddie Albert, Alice Faye, Jan Murray, Phil Harris, and Dean

126

"Oh, this is strong tea. I just melted my teaspoon."

None of us were sure just how that J&B got into the teapot, but our stage manager said he saw Phil pouring something into it just before the cameras rolled. We fashioned Dean and Phil's number after the old comedy team of Gallagher and Sheen, who performed a musical conversation together in vaudeville: *"Oh, Mister Gallagher . . ."* *"Yes, Mr. Sheen . . ."*

The parodies were a hit and we did a new version every time Phil guested on the show, making sure that it was done early in the taping before too much tippling went on in the dressing room. Sober or not, Phil was a great partner for Dean and when they were together the cue cards went flying.

In 1966 Ella Fitzgerald was not quite the jazz legend she became in later years, at least not to Greg's thinking. In fact, he thought she was too fat and probably not at all workable with Dean. Even though Greg and the rest of us were well aware of Ella's unmatched vocal techniques, we were worried about matching her with Dean. Greg's remembrance of Ella apparently went way back to when she recorded "A-Tisket A-Tasket" and he insisted I get her to make

Dean and Ella Fitzgerald

that little relic one of her numbers. "At least she won't be doing any of that — what do you call it? — scat!" exclaimed Greg.

She came to the first rehearsal and couldn't have been more charming or more willing to cooperate, but when I mentioned our desire for her to sing "A-Tisket A-Tasket" I couldn't mistake a dejected frown on her face. Even though she had introduced the song in the '30s and had written it with one of our arrangers, Van Alexander, she begged me to consider something else. "That was a long time ago," she explained.

But Greg stuck to his guns and pushed me into trying one more time. He fought for the familiarity of the song, "an audience pleaser for sure."

Ella relented and began singing the nursery rhyme tune in a very small, even unmelodic voice. The rest of the cast and crew at the rehearsal felt for her. Suddenly she stopped in mid-song and began an old ballad:

You've changed,
That sparkle in your eyes is gone . . .

Our pianist, Geoff Clarkson, knew the song and picked up the accompaniment. A noisy rehearsal grew silent. Everyone was mesmerized. Greg was impressed. "You've Changed" replaced "A-Tisket A-Tasket" and from that moment on Greg said Ella could sing "anything she wants on our show."

I also put her in, of all things, a country medley with Dean and Gordon MacRae. She loved it and did an incredibly wonderful job on "If You Loved Me Half As Much As I Love You." George Gobel, another guest on the same show, felt compelled to join them with his guitar in hand, unannounced, for a swinging version of "You Are My Sunshine." The critics called it our best show of the season.

After it was over, she begged us to have her back soon and said she hadn't had so

Dean, Dinah Shore, and George Burns

much fun in years. Ella was, as always, beautiful in all she did.

When it came to making show business look easy, nobody topped George Burns. Or so people used to say. But when George was a guest on our show he admitted he was a piker compared to Dean.

"I've never seen anyone like him," he told a *Los Angeles Times* reporter. "You're lucky if he turns up the day of the show. And the strange thing is that there's no panic about that around the studio. It's for real. Oh, maybe Dean does take a squint at a script sometime during the day, but you'd better not mention that to NBC or they might fire him. I'm sure Dean didn't just sit back one day and say, 'I'm gonna take things easy.' It's the way he works, that's all. And how can you be a flop if you're not even trying to be a hit?"

Dean with George Burns and Jonathan Winters

We didn't quite know what to do with George.

"I won't take the job if you don't let me sing," he joked. I took him at his word and created a big production number around "It Was a Very Good Year," recreating his radio and vaudeville days with Dean and Dinah Shore. Besides some standards, George threw in a few bars of "I'll Be Coming Back to You, Liza, When I Finish with the Kaiser."

"Hell, you can't go wrong with a song like that," George insisted.

Mr. Kemp had booked Florence Henderson on a show, which brought forth a rousing "Who's she?" from just about everyone on the staff.

"I've worked with her on *The Telephone Hour*," I told Greg. "She's a funny lady. You'll like her."

Greg was ready to take a chance with Florence, determined to make me prove that she could be funny. With Kate Smith on the same bill, it meant that Florence couldn't just stand there and sing. Kate would do that. I thought it might work to poke fun at the sort of thing Florence was associated with, operettas. A capsule all-purpose musical comedy. Greg departed from his ban on special material to let me write the whole thing. Florence camped it up outrageously, falling out of flower-decked swings and tripping up her uniformed lovers.

The silliness ended with Dean coming on the set as her handsome prince (in his tuxedo, of course, but with a bright red sash

Dean and Florence
Henderson

to make the point) and the two of them giggled through some naughty lyrics for
"Makin' Whoopee" and "Just in Time."

Before we taped the number, I hinted to
her that Dean would enjoy some fooling
around, so Florence went all out with jabs in
the stomach, tugs on his leg, even a jump up
into his arms. It surprised and delighted
Dean, who ad-libbed, "Where did *she* come
from? You'll be back, sweetheart!"

Kate Smith was not amused with Florence, especially seeing how much Dean
liked her. She didn't expect this much competition for her host, heretofore her personal property. To more or less make up for
that, I gave Kate two duets with Dean, one
of which included the little waltz section
that she asked for every time she guested.

While Dean was leading her around the

133

floor, she couldn't help resisting a heart-felt feeling of the moment, "This is the bestest!"

Dean did love Kate, even though he couldn't quite control himself when he saw her appear at his orchestra rehearsal in an old housecoat and her hair up in curlers.

"Does she do windows?" he whispered to me. Reviewers continued to consider their meetings on the show "sheer magic," and Dean went happily along with it.

Another "almost regular" was Joey Heatherton, whose bouncy singing and dancing helped our pace and also got us a nod from the younger generation. Contrary to all that boundless energy we saw on the screen, however, Joey would fall into moments of depression from time to time. We began to expect her to be down from the first day of rehearsal. She'd come in and start the conversation by saying she hoped we weren't going to make her do another one of those "jiggle" numbers. She continued to press us into letting her sing a nice, pretty ballad.

Sidney didn't allow her to wallow too long in that. "With the whole world panting to jump on your bones, I don't think so," he'd say. So once again we'd build her the sexiest routine we could possibly get away with.

Dean and Joey
Heatherton

Her moments with Dean, though, took special care. There was always the danger of making their duets look like a *Lolita* affair, so we kept trying to go for laughs. A Mother Goose medley, for instance. On small tricycles yet! No matter, Joey still brought out special meaning to the lyrics of "Mary Had a Little Lamb."

There were rumors about an affair between Joey and Perry Como when she was a regular on his show, but Dean would have none of that. "She's a kid," was all he had to say.

Writer Paul Keyes was a good friend of John Wayne. He did Mr. Kemp a huge favor by personally inviting Duke to guest on our show. There wasn't even the usual movie plug involved. We did have to go along with one request, however. He had just become a father and he wanted to say "some important things" about his daughter Linda's future:

I'd just like to stick around long enough to get her started.

I'd like her to know some of the values we knew as kids. . . . some of the values that too many people think are old-fashioned today.

Most of all, I want her to be grateful . . . as I am every day of my life . . . to live in these United States. I know it may sound corny, but the first thing my daughter's learning from me is the Lord's Prayer, some of the Psalms . . . and I don't care if she never memorizes the Gettysburg Address, just as long as she understands it. And since little girls are seldom called upon to defend their country, she may never have to raise her hand for that oath, but I certainly want her to respect all who do. I guess that's about what I want for my daughter, Dean.

The audience applauded loudly. Fortunately for us all, Duke didn't leave us until Linda was twelve.

Dean and John Wayne

Dean and
the Duke

We had established the pouf, the round love seat, with Dean and some of his female guests, and when Phyllis Diller came on the show she kiddingly asked if she could do a medley with Dean on it, "like all the other sex symbols."

"Wouldn't it be fun if Dean actually had his way with me right there on nationwide television?" she fantasized. So we did it. I put together a bunch of "Lovely" songs that ended with Dean singing:

> *If you were the only girl in the world*
> *And I were the only boy . . .*
> *Nothing!*

Phyllis wrote her own monologues, of

Phyllis Diller

course, but she still wanted to hang around rehearsals, always hoping we might put her into one of those nice, flashy production numbers.

"I long to be glamorous," she pleaded.

"Sorry, Phyllis, but this week you're the comedy and Diahann Carroll's the music."

We'd been warned about Diahann. Those who'd worked with her before called her the "Royal Princess."

"Look out," they said, "she wants to be treated like royalty." We'd been alerted to difficult guests before, but because Dean and his show were flying high we felt nobody could be a real problem anymore. Even the so-called troublemakers were on their best behavior when they did Dean's show. They wanted to be a part of Dean's success and went along with almost anything.

But when Greg called me into his office to apologize for having to go back on his word about "makin' 'em come to you," I knew Diahann was not going to be cooperative.

"She asked if you'd mind driving to her place," he said softly. "I understand she's very, very busy." I assured Greg that I didn't mind at all and would drive out to Beverly Hills with Bob Sidney, her material in hand.

Sidney wasn't pleased about this, but we got to Diahann's home and were immediately escorted to a playroom above the garage. A little rehearsal hall at home, I thought. But after considerable time had passed with nobody paying any attention to us, we figured Miss Carroll was simply too busy to discuss anything about our show, in spite of a confirmed appointment. It didn't help that we could hear laughter and the clinking of glasses from inside the house.

Several hours passed with no word from anyone, and in disgust Sidney and I decided to get out of there. As we drove back to the office we remembered Phyllis's plea to do a production number.

"Let's do it!" we said to each other.

We put together a number with all the

singers and dancers and with Phyllis playing the harp, which she did with surprising flair.

Greg loved the routine, partly because he liked Phyllis so much, but mostly because having her do a big production number instead of the obvious Diahann would be, as he loved to say, "damned unexpected!"

Gail, the third of Dean's seven children, was a striking brunette who I thought was more like Dean than any of his brood. She had his flair, as well as his Italian singer charm.

Gail had graduated from Beverly Hills High, studied theater arts in England, and then returned to tackle the nightclub circuit. So far, she and her older sister Claudia were the only ones of Dean's offspring so inclined. Dean had been cool about it, merely advising Gail that if she took too many lessons she might wind up in the Mormon Tabernacle Choir, holding her hands.

"Play it loose, honey," was his only advice.

When Gail's name came up on the show's guest list we all thought Daddy had made a phone call. Not so. Greg had figured she had gotten enough publicity and decent reviews to warrant a guest shot.

"The public will be interested," he said.

Gail was considerably less nervous than

we thought she would be. It was obvious a lot of the old man had rubbed off on her. "I was rehearsing in my room at home," she told me one day, "singing at the top of my voice. I used to think loud was good, you know. Anyway, Dad rushed in and yelled, 'What in the hell are you shouting for? You got the job!' I got the point."

Gail knew from talk at home how the show operated. Everybody rehearsed but her father. She wanted as much rehearsal as she could get. The audience pleaser in her nightclub act was a rousing rendition of "Rose of Washington Square," so Sidney and I created a production number around it. Gail had made a point of perfecting what dancing she'd been exposed to along the way, and it was, to everyone's delight — es-

Dean and daughter Gail

pecially Dad's — a pro number.

I followed it with a duet with Dean, in which both of them showed a lot of love and admiration. We knew the fan mail would come pouring in, and it did.

Dean seemed genuinely happy that his daughter had been on the show and had performed so well. It was also the first time his wife, Jeanne, called Greg. She wanted to thank us for taking such good care of her little girl. Because any word from Jeanne was so rare, we treasured that call.

My respect for Mr. Kemp's booking ability was boosted considerably when I discovered he'd lined up Caterina Valente for an upcoming show. Even though she'd been a semiregular on *The Danny Kaye Show* and I had worked with her when she was a costar on *The Entertainers,* she was still not well known to American audiences, and therefore practically a nobody to both Dean and Greg. But I knew how extraordinarily talented she was and I figured that everybody on our show would see that as soon as she began rehearsals.

She could sing, dance, play the guitar, almost anything asked of her — and do it extremely well. Her English had only the slightest accent, just enough to make it

Caterina Valente, Dean, and Dom DeLuise

charmingly cosmopolitan. A production number was no problem; she did everything full-out and exquisitely. But to introduce her to Dean, I decided to have her accompany him on the guitar, which would be just personal enough to make him like her a lot, which he immediately did. She taught him the "One Note Samba." All he had to learn

Dean and Caterina Valente

was one note. As the cameras were fading away at the end, she joyfully exclaimed, "You did it! You did it!"

"What did I do?" Dean asked, confused.

"You sang, 'ba-a-a-a,' your one note. It was beautiful."

"You bet your sweet cheeks!"

Dean was a goner. Instead of fading to black we let the cameras roll as he swept her off her feet and carried her off the stage. He was stunned by her talent and didn't hesitate to say so, on camera and off. She'd be back!

When I was growing up in Tacoma, Washington, and going to seemingly every movie ever released, my "sweetheart" was Alice Faye. I had this thing about 20th Century-Fox movies in general. There was something different about them; a look, a sound, a feeling. Alice, in those days, was their big musical star and I saw every movie she made, over and over. She was so believably nice up there on the silver screen, with that dulcet husky voice that could tear your heart out.

Naturally I was thrilled beyond belief that she was about to be a guest on our show. I'd write the most fantastic musical material for her! This would be the absolute apex of my career — meeting, writing for, and singing with *Alice Faye!* I couldn't wait.

I went to work on a long spot, a musical summary of her entire life, lovingly created for my dream girl. "She's on her way to your office! She's on her way to your office!" warned the staff (they knew how I felt). I was a basket case when she walked in and introduced herself. The years had been good to her. Her face was still big, round, and pretty, the figure sensational.

She was immediately everything I had hoped and dreamed of.

Alice Faye

"Greg tells me you're going to help me get through this show," she murmured. My God, even the famous lower lip quivered, just as I remembered.

"Nothing could surpass this!" I thought. "I might as well stop my career right now!" I couldn't wait to present my musical declaration of love. She'd be so thrilled with it, she'd hug and kiss me, expressing great gratitude and admiration. I sang it down for her. My notes were shaky; my nervousness must have showed.

It was about a six-minute extravaganza, and when it was over I tried not to show that I was totally exhausted with anticipation. I waited for her to declare it to be magnificent and overwhelming.

"It's lovely," she managed, after an awfully long wait. Lovely? That's not quite the word I was hoping for. Not for this, the ultimate in testimonials to 20th Century-Fox's musical queen. Lovely?

She was too shy, I thought, to accept anything that glorified her so outrageously. Or worse — did she think it was the worst piece of crap she'd ever heard?

Either way, I struggled through a most embarrassing silence. I sensed my tribute was not the great piece of work I'd imagined it was and quickly suggested a medley of her

147

movie songs might be more appropriate.

She seemed much more at ease with that and asked which ones I felt would be best. I knew them all, so I blurted out several of my personal favorites and she suddenly beamed with satisfaction.

"I'm in your hands. Whatever you select is fine with me." Well, at least I was choosing Alice Faye's songs, even if she wasn't going to sing my special material.

As a devoted fan, I'd read that she was uncomfortable about running through rehearsals with staff and network brass around, so I talked Greg into letting Sidney and me go through her production number and do her singing and dancing during the show's first run-through. She greatly appreciated that, but when the audience came in and the big lights went on, she was, to me, still the greatest star!

After the taping she invited Sidney and me to a nearby restaurant for dinner. Her husband, Phil Harris, would be there, too. Mr. Kemp had booked them on the same show. After a couple of stiff drinks she confided in me.

"Everybody told me how you felt about me when I first walked in," she said. "You know something, I've been a wreck all week trying to live up to it! You don't know what

148

you put me through. But I thank you very, very much."

I apologized and told her I just couldn't help it.

"Now that it's over," she admitted, "it was great fun. I've never had so much attention."

She began to go on and on about how absolutely wonderful the entire staff had been to her, glowingly elaborating on each and every stagehand, musician, and dancer until Phil stopped her in mid-adjective. "For God's sake, Alice, cool it. Have another drink!"

After her third Scotch Alice became more candid. "I didn't care too much for some of those steps Sidney made me do, though." Sidney smiled.

Another Scotch and water. "Those dancers are a bunch of fuckin' drags, aren't they? Who was that blonde bitch?"

I mentioned a few names and she added that they were *all* bitches.

"Maybe we should order dinner," Phil said.

"And that rat Roberta. She just had to tell everyone I was on the wrong fuckin' foot, didn't she?"

"I think I'm going to be sick," I whispered to Sidney.

"Don't be silly," he said. "She's just letting off a little steam."

I was still seeing Alice as that sweet young thing on the screen. But this was the real Alice. Imagine that! She was just like everyone else.

"You're quite a gal," I said to her, "and thank you for being the old Alice for me all week."

"What do you mean, the *old* Alice!" Suddenly she got up on the banquette and began doing some of Sidney's steps from the number we'd just taped.

"I'd like to see Cyd Charisse do that!"

Alice made many more visits to our show and never mentioned my musical salute. I tore it up one day. Whenever she appeared, I put together another medley of her film hits (there were many) and I'd go through the rehearsals glowing in the delight of working with the wonderful Alice Faye. She was still my dream girl.

Dean was beginning to acquire a stable of lady guests who worked well with him, but perhaps no one met all the requirements as well as Petula Clark. The chemistry, as they say, was right.

Her visits were always a treat for me because there seemed to be nothing musical

Dean, Alice Faye, and Van Johnson

she couldn't tackle — and fast. I'd play a medley for her once and it was instant memory. Rehearsals were short because she would master everything quickly and there was no reason to repeat and repeat.

There were no key problems, and she seemed to have a limitless range. We'd trot out the old pouf and Sidney would say, "Start here, move to the front on the second song, sit on the third — after that you're on your own." She'd simply take Dean's hand at the beginning of their duet, go where she was told to go at the proper time, push him down on the pouf when need be, and pull him up when she felt the tempo change called for it. It was all done on impulse, yet

151

looked as though it had been very carefully choreographed.

"She looks a little like Dean's wife, Jeanne, don't you think?" people would say. Maybe that's why there was all that chemistry. Something obviously was going on between them, if only for the ten minutes or so they were on-camera. There were rumors about them being caught wildly embracing in her dressing room, but for us it was what was on-camera that counted — chemistry that couldn't be directed. Greg would see the eye contact and go in for close-ups.

I encouraged it by putting into each medley a tender ballad, some song Dean would know so well that he wouldn't have to turn away to the cue cards for the lyrics:

I give to you and you give to me,
True Love, True Love.

Did anything really happen between them off-camera? If not, Petula was an awfully good actress. It was pure magic. Whenever she was on the show it always seemed that next week's lady singer was so anticlimatic that nothing would work. We just couldn't top the Dean-Petula combination. I'd struggle with ideas to use for the other women and was almost always disappointed. Petula spoiled us.

Connie Francis, for instance. She had retired from show business several years earlier and here she was the week after Petula, obviously trying to pick up the pieces of her career.

We wanted to help. Her first request, though, was that we wouldn't ask her to do any of her old hits. She wanted to do something new and fresh. We ran over several from the current Top 40. "Depressing," she'd say.

The miniskirt craze had just blossomed and I wondered aloud if anyone had written a song about it yet. They hadn't and Connie thought it was a great idea. Van Alexander

Dean and the Mills Brothers

153

Dean and Petula Clark

and I wrote one for her called "Cute Little Mini Skirt." She loved it and expressed a desire to record it, but as far as I know, never did. I don't think she could interest any record companies then. Since we had just done the pouf bit with Petula, I searched for another approach for the Dean and Connie duet.

I couldn't picture them being romantic together so I went to the opposite extreme, a patriotic medley with lots of state and city songs, ending with (what else?) "God Bless America." To everybody's surprise, the studio audience gave it a standing ovation. Even Dean was startled. You can't miss with "God Bless America."

There were other firsts that season.

Dean's famous fireman's pole was one of them. Sliding down it at the start of every show became one of his trademarks. In fact, of all the crazy surprises we threw at him, he remembered one week's slide down the pole as the most harrowing thing that ever happened to him during the series.

"Some stagehand, thinking he was doing me a favor, rubbed it down with some greasy substance and I sailed down so fast I thought I would go through the floor. The bottoms of my feet hurt for weeks afterwards."

There were more station break teases this year, too, almost always with Dean and the girls. One memorable one had him fitted with a specialty vest with holes connected to tubes of water, the flow of which was controlled by someone offstage.

It began with the girls putting Dean, cocktail in hand, in a huge box; then the girls supposedly inserted large swords through the box, and into Dean. As the music ended Dean walked out of the box while the offstage controller released the water, making it look as if Dean had been pierced with the swords and was spurting fountains. Dean would try to read the cue cards for the tease but each time he began talking, the controller would pump out another gush of water.

Completely at the mercy of the guy off-stage, Dean did his best before resorting to some bleepable words not on the cards. Naturally, it was all kept in the final edited show.

Bing Crosby showed up for our last show of the season, and he and Dean seemed more relaxed and casual than ten Perry Comos. It was often reported that Dean had copied Bing's style when he was getting started (perhaps it was more Bing's attitude). Dean admitted that, rather than Crosby, his phrasing came from watching Harry Mills of the Mills Brothers. In the times the brothers were on the show it was easy to see Dean doing a good imitation of Harry and vice versa.

For Bing I suggested a medley of his hits. He said he'd rather do that with Dean and not do a solo himself. Greg agreed that since we had two *really* big stars the medley could be nice and long. Since they were all Bing's songs, we didn't have to rehearse much, and Bing didn't ask for any changes. He was the ultimate professional — memorized everything he had to do, didn't complain about anything, and disappeared as soon as possible.

Dean and his fireman's pole

The most impressive thing about Bing was the volume he got out of those mellow low tones. They were downright loud. And of course there was his well-known easy-going attitude, apparently the same off-screen as on.

The duet went well. The notes were all there along with the super-casual style of the two stars, yet I felt it was almost so relaxed that it seemed too long, as though they just might fall asleep if there were one more song.

Bing was out the door before Dean, never saying good-bye to any of us, not even a "thanks for the medley" or "mention my name in Tacoma." (His hometown as well as mine.)

Earlier that week he had asked rehearsal pianist Geoff Clarkson to sketch out a similar medley that he could perform the following week on the road. Geoff took considerable time putting it together to Bing's liking and discreetly refrained from asking for any money, knowing Bing's reputation as a miser. Geoff was surprised, however, when Bing said he wanted to give him something for all his hard work and presented him with a tiny bottle of Scotch!

Even at Christmas, Bing's magnanimity raised eyebrows. His little gifts to staff and

Dean and Eddie Albert

those around him were always tax-deductible, and whatever they were (usually an inexpensive ballpoint pen) you could find the words "Bing Crosby Enterprises" stamped on them.

Bing was professional, loose, and witty — but he sure flunked generosity!

We finished the season sometime in March and plans were begun for another summer replacement show, *Dean Martin Presents.* . . . This year, however, Greg decided to let someone else produce it and he wouldn't let me take part either.

"Give yourself a rest," he said. With that

he presented me with a bonus and a round-trip ticket to Europe. He was always up front: "This is a bribe not to do a summer show for anyone else."

Before I left, there was a wire from the Television Academy announcing that *The Dean Martin Show* was nominated for an Emmy for the previous season. Three Emmys, in fact. One for Dean and the show itself as "Best Variety Series," one for Greg as "Best Director, Variety Series," and a third for "Best Writing, Variety Series."

Around the tenth show of the second season, Greg had decided that since I was writing all the introductions to my medleys, I deserved being listed with the other writers on the show's credits. I had previously gotten "Special Musical Material by . . ." and this meant that my name would come up two times on the crawl.

I quickly joined the Writer's Guild and my name appeared twice several weeks in a row before the other writers objected to my dual credits. They won their argument with Greg and I was pulled off the list of writers. But the academy included me and I got an invitation to the annual ceremony at the Century Plaza Hotel.

Dean didn't normally go to industry affairs, but he said he'd make an exception

Sid Caesar, Dean, and Bob Newhart

this time because he was sure we'd win. Greg frowned on these things, too, but this year seemed different, and we all went. Even Jeanne insisted on attending. Like the rest of us, she felt we were a shoo-in.

Yet we all wondered if our assigned table (at the extreme back of the room) was any indication of how we'd fare in the voting. When the "Best Variety Show" category came up, it seemed the entire audience turned toward us, as though they, too, expected us to win easily. The place grew quiet and the winner was announced.

"The Andy Williams Show."

Jeanne, disgusted, was the first to storm out. Dean wasn't far behind, mumbling something about this being the "last damned time I'll ever show up for one of these stupid things." He was good on his word.

In later seasons, the show itself, the writers, and Greg and I garnered many nominations but no wins. However, Dean wasn't the least bit interested in the Emmys, win or lose. "The food is lousy and you can't get a drink unless you've got a steamroller to get through the crowd to the bar."

4 The Third Year — (1967–1968)

More and More

Right after the Fourth of July holiday in 1967 we began our third season. Greg wanted to start early in order to get as many shows as possible "in the can" before the premiere air date in September. The first show on the air would not be the first one we taped. He wanted Dean and the rest of us to "settle in," so several down-the-line shows were taped and put away for later in the season.

Back in the halls of Burbank, our first taped show starred the wonderful Rosemary Clooney. Rosie had just gone through a serious breakdown, plus a traumatic divorce from actor Jose Ferrer.

"Be especially nice to her," was Greg's requirement. That was easy. When she arrived for the first rehearsal, Rosie was in fine shape — physically, mentally, and vocally. We didn't discuss her problems, of course, but we were all pleased that she was

Rosemary Clooney, Dean, Minnie Pearl, and Buddy Hackett

in super form again.

Greg also told me that he knew Dean was very fond of Rosie and that I should make their medley together reasonably long. I whipped up one in which they each sang out the other's praises: "If You Knew Dino Like I Know Dino," "Rosie, You Are My Posie," "Whatever Rosie Wants, Rosie Gets . . ."

Greg thought it was so good that he upped my salary again.

Dean and Rosie hugged their way through the medley, planting sincere kisses on each other at the end, and Rosie was obviously pleased.

I put the two of them in a long finale as well, a bit of special material by Leslie

Perrin I always loved. It included Minnie Pearl and Buddy Hackett. Called "Cook's Tour," it was based on the fact that every time vacationers go abroad, they bump into the same tourists they've met previously, including Hymie Licktenberger (Buddy) from the Bronx.

In the middle of the number, Buddy suddenly ad-libbed: "In Spain, my wife did something nobody has ever done at a bullfight. She threw up on the man in front of her." We had to wait for Dean, Rosie, Minnie, and the audience to regain control. Unfortunately, Standards and Practices made us bleep the line.

What was to be the first show had to get the most attention. It was the only one to get reviewed and the industry would pay heed to whatever ratings we got. An all-star cast was booked: Jimmy Stewart, Juliet Prowse, and in his first variety show ever, Orson Welles.

We all had the jitters about Orson. His reputation for greatness in every other field of show business was universally accepted and we didn't want to make fools of ourselves. Along with the rest of the world, we admired Orson. He went against the rules, a lot like Dean. He was able to do what everyone said couldn't be done. Someone once said that if Orson hadn't been born, he surely would have invented himself.

One thing he couldn't do, though, was drive. Greg usually picked him up at the Beverly Hills Hotel to take him to the studio. When he couldn't, I was given that honor.

"Whatever made you agree to do an everyday TV variety show like ours?" I asked him on one of those trips.

"The money!" he answered without hesitation, in that booming voice.

But that wasn't entirely true. When Greg was eighteen he met Orson in Chicago. Orson had three daughters from three different wives and then, he said, "I had Greg." He put Greg through school. So the reason Orson did our show so many times was because of Greg, although neither one of them ever mentioned that.

"Dean Martin fascinates me," Orson would say. "How does he get away with it? He's a maverick, like I was. He makes incredible mistakes and everybody loves him for it. But he knows exactly what he's doing and he's getting away with it! I like that. I admire him tremendously for it."

When I brought up the subject of his getting away with imitating William Randolph

Dean, Jimmy Stewart, Orson Welles, and Greg at orchestra rehearsal

Hearst's life in *Citizen Kane*, he squirmed a bit and I felt that "Oh, no, not that old subject again" feeling.

"Well, it wasn't based entirely on him, you know. The coincidences in the story made everybody think so."

"Sure," I said to myself.

"It gave us a heck of a lot of publicity, didn't it?" Orson added. "I have to laugh now, but even when that notorious Hearst columnist Louella Parsons visited the set, she didn't have the slightest idea of what we were doing. She really was dumb, so we let her hang around. When the picture was finished, she was furious. Actually someone in our script department gave a copy of the screenplay to Marion Davies and all hell broke loose."

Dean and Orson Welles

That having been said, he got back to the subject at hand — doing our show. He couldn't get over the phenomenon of Dean Martin. "Imagine coming out completely unprepared and making it look so easy. The guy's got it."

I asked if he might try doing the same thing.

"Not a chance. I'd be a fool to try. So would anyone else. Dean's one of a kind, the only person I know who could make a success of it. God knows the rest of us need all the preparation we can get."

He wanted to know what we had planned for him. I told him Greg wanted him to do a Shakespearean reading of some sort, his choice of course.

"Seems a bit odd in the middle of a *Dean*

Martin Show, doesn't it? But I suppose it does come under the heading of variety."

I added that I thought it would give us a bit of class. When he got to the studio, he found he'd be doing much, much more, including a little personal story about his early days in radio — which he totally rewrote. The show had a sketch about a radio mystery show in which Orson was the narrator and Dean the bumbling sound-effects man dashing all over the studio, three choruses of "Brush Up Your Shakespeare" with Dean, and another sketch with Dean and Jimmy Stewart under hair dryers in a men's hairdressing salon that led into a minifinale with the three of them doing a song and dance.

With his soft talk and stammering delivery, Jimmy seemed just the opposite of Orson. But both were eager for the challenge of nonsense and fun ahead. Besides the sketch and song with Orson, Jimmy did a bit of patter with Dean in the "what stars are really like" mode, ending with his famous piano rendition of "Ragtime Cowboy Joe," the only song he knew how to play — and sing.

He was thoroughly fascinated by the way we put on a show and wanted to hang around even when he was not involved with rehearsals. When we gave him a call, he'd ask what time the others were arriving and show up at the same time they did, if only to sit around and watch.

I relished the opportunity of working with these two movie giants, standing in for their costar Dean, and cueing them in whenever they needed it.

I pushed and pulled them around all week, explaining that they'd have to know what they were doing because Dean wouldn't. "You'll be on your own when the lights go on," I reminded them, and they didn't mind practicing it over and over until they felt they had perfected it. Neither one had been involved with special lyrics and dance steps before, so we made it as simple as we could. When it was time for taping, Jimmy became the leader, reminding Orson and showing Dean how it was done.

They even agreed to don hippie wigs at the end and successfully made it through a short minuet as well, even though Sidney and I had to stand in front of them off-camera to direct. We'd do the steps and the three of them would follow us.

Orson picked Shylock's speech from *The Merchant of Venice* as his nod to "class," but suggested that Dean's audience might need some sort of explanation. "A lesson in bigotry," he told them, "for those not familiar with the classic."

Juliet Prowse was a sexy singer-dancer, which we took advantage of in a strong solo number called "The Very Soft Shoes." Orson commented on the "drama" she put into it, noting that she was the most talented dancer he'd ever seen.

Dean and Juliet Prowse

It was one of many visits to the show for Juliet, who knew how to handle Dean right from the beginning. She took him through a sketch about warming up at dancing school.

"We start by going right up to the barre," she said.

"Hey, I'm for that."

"Now put your foot in second position. You know your positions, don't you, Dean?"

After some feeble attempt at leg stretches and pliés, Dean ad-libbed, "I think I got my shorts on backwards."

"Come on now, deeper . . . deeper . . ."

"I just hurt my tutu!"

As usual, it couldn't have been that much fun if it had been rehearsed. At the end of a rousing polka, which Dean only did once, Juliet jumped into his arms. She said she just felt like doing it. Dean loved it.

During rehearsals, I had to convince her that she must be strong with Dean, that she had this one chance, and that Dean would follow anything she did. Once again, the same pattern — the stars leading the star. "You have to be a dance instructor when we get on camera," I told her. She was indeed strong with Dean and he reacted as predicted.

We were off to a rousing start. Jack Gould in the *New York Times* reviewed that first show:

Orson Welles, the corpulent expatriate of multiple theatrical talents, made his TV variety bow Thursday night as one of Dean Martin's guests. He could not have been in safer hands. While Mr. Welles recalled the days of radio soap opera, Mr. Martin was the hardworking soundman of yesteryear. The quick gag had its fun because Mr. Martin held the routine together without impinging on Mr. Welles' function as a star.

In fact, Mr. Martin for some months now has been a thoroughly agreeable phenomenon of TV show business.

Thanks to his naturalness and winning disdain for excessive rehearsals, he achieves a quality of engaging informality that is carrying him far. He pleasantly shares the spotlight with others and manages to take the curse off the usual Hollywood cues, as he demonstrated last night by drifting into a comic sketch and duet with Jimmy Stewart.

Mr. Martin's gift for likeable self-deprecation is providing an air of spontaneity to supplement his singing style. With Greg Garrison's help he has intuitively found one of the secrets of TV longevity: he doesn't take himself seriously, one of the hallmarks of professionalism.

At the end of the first show Dean added a little personal touch to his good-night, something he strongly wanted to say: "Tune in to Jerry Lewis's new show Tuesday night . . . I'm going to." His ex-partner was taping his new weekly series right across the hall

Don Rickles's first network TV appearance

from us. Jerry, of course, hung around NBC all week, rehearsing his show to a fare-thee-well. Dean was to show up on Sunday and the whole studio waited to see what would happen when he walked in. Would the two performers, who had supposedly not spoken since the breakup, meet at last? Their dressing rooms were side by side. Their schedules seemed dangerously identical. Each of them had said unkind words about the other in recent years. There was a good chance that something explosive might take place at NBC. But aside from someone hanging a sign on the back of Jerry's Rolls-Royce that said, "Ban the Bomb Before It Goes On," everything was quiet. The tension never developed. Thanks to some careful planning they never ran into each other. Jerry's show went off the air after thirteen weeks, his third try and his third flop, while Dean's show continued to reach even greater heights.

We started to appear regularly in the top ten of the Nielsen ratings. No. 7, No. 5, No. 2 . . . close but no Kewpie doll yet. Mr. Kemp bowed out of the picture for health reasons, and Greg took over the booking. It meant that there would be a maximum of four guests per show, each chosen with a selective eye to make the most of their moments with Dean.

"I don't care about the production numbers. I don't care about the comedian's monologue. I don't care about the guest singer's solo." Greg was showing his new

Dean and Ricardo Montalban breaking up at Don Rickles's jokes

Patrice Munsel, Susan Barrett, Dean, and Van Johnson

power. "It's Dean that counts. It's Dean they talk about every Friday morning."

The guests became objects that were used to highlight Dean. Still, Greg recognized the importance of "pace," and if Petula Clark was going to be on the show, everyone would expect a solo from her. "Not two or three songs," Greg added, "just one number and a long medley with Dean."

Petula's show, our third for the season but first to be taped, had Flip Wilson and a guy Greg found in the lounges in Las Vegas, a wild, insulting comic by the name of Don Rickles. Greg felt that television audiences deserved to "discover" Don the same way he did, in a nightclub atmosphere. He turned our entire Stage 4 into a Vegas club, by inviting as many stars as he could find to sit in the audience. He dragged them out of the halls of NBC, got them off the golf course, went searching around nearby movie studios, and made a few begging personal calls of his own.

Dean set it up: "You know, Las Vegas is becoming the entertainment capital of the world. Practically every great act in show business works there at one time or another, and when they do, every night after their performance they run in to see my next guest. He's got his own style of comedy. He gets laughs insulting stars who come to see him. And they must love it because they keep coming back. I went to see him recently and in the audience were more stars than an Emmy show. Rather than tell you what happened that night, I'll show you . . ."

In our studio was an impressive bunch of biggies: Danny Thomas, Lena Horne, the Andrews Sisters, Polly Bergen, Don Adams, Jackie Cooper, Ernie Borgnine, Bob Newhart, Ricardo Montalban, Macdonald Carey, Pat Boone, Joey Heatherton, Barbara Eden, Ross Martin, Caterina Valente, Pat Collins, Rose Marie, and a last-minute walk-on by Bob Hope. Don quickly let a few zingers fly.

To Pat Boone, who toasted Don with a glass of milk: "The pimples, you still think they come from Hershey bars, eh, Pat?"

To Ernie Borgnine: "You were so great in *Marty*. Now it's over."

When Bob Hope finally arrived: "Why is he here? Is the war over?"

Greg let the section run for a good hour in the studio, from which he picked a rare 21-minute chunk for the final edited show.

It was a turning point in Rickles's career. Up to then he was a club comic that show business personalities talked about but TV producers were afraid of. They thought he was too much for the rest of the country.

Van Johnson and the girls

"Greg had the courage to use me when nobody else would," Don said later. "They were all scared of me. I couldn't get arrested on *The Tonight Show*."

Greg's gutsy gamble paid off. After Don's appearance on our show, everybody wanted

Me with Dean, Greg, Frank Sinatra, Dino Martin, and Frank Sinatra Jr. at orchestra rehearsal

him. We put Don in a western sketch in another show later in the season. Sidney rehearsed Don's part all week and I played Dean. That left only Roy Rogers and Dale Evans knowing their lines in advance. When the taping came, Dean and Don simply winged it, leaving precious little of the original sketch intact. Roy played the stalwart western hero, Dean the outlaw, Dale the rowdy saloon singer, and Don the bartender, constantly offering Dean a shot of whiskey which he was to unhappily refuse.

Greg told Don he didn't care what he did in the sketch, to rip it up if he wanted to, which of course he did. A pussycat off-camera, Don's insults and total disdain for

179

written sketch material only improved what might have been a mildly funny scene at best. "Don't worry," he told Roy and Dale before we taped, "it's all in fun. I won't hurt you."

If Don tore the place apart in this sketch — which went on for at least twice its scheduled length — the rest of the show was handled by Roy and Dale. Their best routine was a hoedown using the entire cast. Dean was literally thrown in at the last minute, pushed and pulled through a square dance in which he didn't have the slightest idea where he was going. The cameras followed his confusion closely and everybody had a good time.

With Don and Dean never around, the week seemed interminably long. Sidney and I threw all the singers and dancers into Roy and Dale's numbers just to pump some life into the two western legends. Dale spent most of the time in a corner sewing and Roy was forever scanning a racing form. We'd get them on their feet from time to time to show them where they'd have to stand, then release them to their respective sidelines as soon as possible.

The fact that we had so little to do that week prompted Greg to wonder why we couldn't do *two* shows in one week. "We could bring Dean in on Thursday," he said. "He's not making a movie now." After trying a few Thursday and Sunday tapings, Greg decided it would be just as easy to rehearse two shows a week at the same time and tape them both back-to-back on Sunday. "Maybe Dean could come in a little earlier on Sunday," he suggested.

He decided to wait to put his new plan into effect, however, until after the Christmas show, which would be a nice, family outing, with just Dean, his wife Jeanne, and their kids. Oh, we'd have Frank Sinatra and his clan too.

Patrice Munsel, the opera star, was on a show with Van Johnson and pop singer Susan Barrett. Pat said she didn't want to do anything operatic, but I was hoping she would so that it wouldn't conflict with a Broadway-style number we had planned for Susan.

I compromised by giving Pat a Gershwin waltz called "By Strauss," in which she would have a little flair of Viennese soprano passages as she went along with the Gershwin tune. Producing it with singers and dancers and having Pat run all over the stage while singing at the top of her voice, I hoped that perhaps Greg would allow her to take a breath before her song with Dean. He promised to do that, but when she finished "By Strauss," he sent Dean right out on stage to join her. Pat went graciously along with it, huffing and puffing and allowing Greg to smile broadly because the show looked live. Dean and Pat's song was "Makin' Whoopee." It wasn't staged; they were just going to sit and sing it straight. But after the first eight bars, Dean began to play patty cake after every "we're makin' whoopee" line. Pat went happily along with it, and it got them giggling so much that they missed a couple of entrances. But Greg was thrilled, and it stayed in the tape.

Van Johnson, in his trademark red socks, was another frequent guest and loved

Dean and Frank

working with Dean. Knowing that he started his career as a Broadway chorus dancer, we put him in many production numbers with the girls and always a man-to-man songfest with Dean. They obviously enjoyed each other.

That show's finale had Dean, Patrice, Susan, and Van in a twenties collegiate musical. Dean, of course, was the dumb football player ("I used to play tight end a lot," was one of Dean's ad-libs) who had to pass a test before he could play in the big game. The determining final question was in mathematics. "How many ounces are there in a quart?" It didn't take Dean long to answer. "Thirty-two!" He passed the test and everybody danced a wild varsity drag.

When we began planning the Christmas show, Gail Martin had already been on and knew how we worked, but the rest of Dean's family was an unknown quantity to us as far as music was concerned. Dino had been making rock records with Desi Arnaz Jr., so I was aware that there must be some ability there. Deana made a special trip into the office to assure us that she could perform as well. As for Craig, the eldest, Greg had put him on as production assistant the previous year and he had done well enough to be promoted to associate producer when Norm Hopps left for *Laugh-In*. He made it perfectly clear that he was not in any way, shape, or form a performer like his dad.

"I've got a tin ear, and don't you forget it!" he'd yell whenever we suggested a turn with his father. Claudia had done some acting in several forgotten movies but in-

(*Clockwise from top left*) Craig Martin, Dean and Gail Martin, Dino Martin, Claudia and Deana Martin

sisted we couldn't count on her for anything musical. Dean's younger children, Gina and Ricci, were just kids and we knew we'd do well just to get them to sit down and be patient during the lighting and camera-angle delays.

The Sinatras presented no problems. Nancy and Frank Jr. were established singers and Tina was known to be okay in that department as well. The current Mrs. Sinatra, Mia Farrow, didn't show up. There were rumors that she was sitting in a secret dressing room somewhere at NBC watching everything we were doing.

Jeanne, who usually never came to a taping, agreed to be in the closing "serious" spot with the rest of the Martins and Sinatras. She had told a reporter she preferred not to come to the studio. "He's worked hard all these years," she said, "and I'd rather share this job with him at home, in front of a TV set on Thursday nights."

We knew we'd have the kids all week and that Frank would probably show up the same time as Dean — at the last moment. We concocted a show that let the kids do most of the work, in duets or trios, ending with an appearance from the two dads. The viewing audience wouldn't care about the kids' individual talent, but they would want

Dean and
Dino Martin

to see how they looked and how they performed with their illustrious parents.

I put together a ten-minute medley for Dean and Frank to do by themselves. Introduced by their daughters ("Ladies and gentlemen, my dad . . .") they swung through some fifteen standards and the result was a classic TV moment. There was no scenery, no direction, no rehearsal. The cue cards were all they needed.

Time for the taping came and the kids couldn't wait to show Dad what they'd been up to all week. But Dean and Frank surprised them by not even going through camera rehearsal with them. Instead our choral director Jack Halloran and I con-

186

tinued our stand-in roles down to the final seconds.

To avoid confusion on the set, Jack wore a big card reading "FRANK SINATRA" on his chest, and I in turn wore one with "DEAN MARTIN" on it. Later, just as Dean and Frank were called out to start the show they appeared with the same-sized cards on their respective chests reading "JACK HALLORAN" and "LEE HALE."

The closing spot, the only serious moment of the show, featured the entire cast singing carols and seasonal ballads, even Jeanne lip-syncing "Silent Night."

The reviews were glowing.

The *Hollywood Reporter:*
What might have been a sweet, sappy hour with another group was instead a raggedy-Andy, rather chaotic Christmas outing with the Pack. The Martin and Sinatra kids all came off showing a great deal of finesse.

Variety:
Frank and Dean acted like Katzenjammer Kids and for home sitters that must have suited them fine after a fairly steady diet of serious productions. It was never irreverent. Garrison and his staff rate a wreath for what will pass as one of the season's best.

Dean with Jeanne and their youngest, Ricci and Gina

Ben Gross in the *New York Daily News*: *Quite often some reader of this column asks the question: "Do you really think that the children of such stars as Sinatra and Martin could have broken into show business as easily if they hadn't had such famous fathers?" The answer is No. It can't be denied that doors are easily opened for such fortunate kids. But once they enter these portals they have to make good on their own. The Sinatra and Martin younger generation is certainly carrying on the traditions of their elders.*

Hal Humphrey in the *Los Angeles Times*: *Watching Martin and Sinatra work this way gives the casual observer the impression they aren't working very hard. To some extent that may be true, but it also is a mark of their professionalism. They don't have to worry about myriad details of performance anymore because experience has made those things second nature to them now.*

Best of all, the show shot up to No. 1 in the ratings, the highest rated show of the year.

Greg felt it was time for a full-page ad in *Variety*:

WHY ARE WE NO. 1?
We try harder, that's why.
 You'll never find a dirty ash tray in our dressing room. We say "please" and "thank you" to our card boys. You can pick us up on any NBC station, anywhere in the country. We don't take Sundays off. Our girls look like girls. Mess with the controls and you can get us in any color. We love everybody. Especially No. 2!

DINO

(I've always loved everybody.)

Joel Grey is an extremely talented performer and we were excited about having him on the show. He and Sidney mostly fashioned his solo dance routine, but when it came to what he would do with Dean, Joel and I had a few differences. He didn't like the song I'd picked, "Way Back Home." He especially hated the lyrics I'd written for it about L.A.:

The doctors the healthiest
The dentists the wealthiest
The love-ins the hippiest
The nightclubs the clippiest
The days the clammiest
The traffic jammiest . . .

And then there was the key, which, as in all these cases, was in Dean's favor. He asked for it to be higher.

Almost every other guest knew the value of that moment with Dean and they understood that things had to be geared to that infamous last-minute appearance he made. But Joel argued that *he* had to look good, too. I suppose he was right, but Greg had taught us all that Dean came first, and if anybody didn't like it, he would gladly escort them off the show. I finally convinced Joel that what I had written for the two of them would work because it would be easy

Dean and Lena Horne

The Andrews Sisters, me, and Lena Horne in rehearsal

for Dean and therefore enjoyable to all.

Oddly enough, although he regretted the material, he ended up asking for more time with Dean. Their bit together didn't really work, not especially because of the material, but because there was very little happy communication between them. Once again, Dean couldn't warm up to what he considered "Broadway slickness." Joel's solo number was top-notch, however.

Lena Horne was on the next show and for the first time she and the Andrews Sisters were able to get Dean out of his tux for a tramp routine.

Lena was one of our top guests. She

Dean and
Cyd Charisse

seemed to be better than ever these days, a
hard act to follow. Greg asked Sidney and
me to produce one of her songs, complete
with singers and dancers, each time she appeared on the show. This time, because he
had seen her dance up a storm somewhere
(probably on Broadway in *Jamaica*), he
wanted her in something low-down and
sultry.

"Get her to really move her ass!" he insisted.

"Not at these prices," she shot back at us
every time we made the pitch. Lena was the
first to ask for $10,000 a show, well above
the established $7,500 that all variety shows
gave their biggest guests. Sidney and I
thought she was well worth it — and so,
down deep, did Greg. But of course he

Dean and
Melissa Stafford

couldn't let anybody get away with that.
The pattern would be broken and every star
would get greedy. Besides, as Greg constantly reminded us, "Dean is the show; the
guests are secondary." Lena eventually
agreed to the lower fee but not to the dance
number. Nobody ever brought up either
subject again.

The salary problem came up again with
Woody Allen. Greg liked him but felt he
didn't deserve even the $7,500 at the time.
Woody agreed to come by and do "whatever
I feel like doing," adding that he didn't want
anything scripted and would not work in
musical numbers or appear with Dean.
"We'll tape what he does," Greg said, "but
we probably won't air it." Woody went on

194

for well over ten minutes and Greg thought it was so funny that he left the entire routine in the final edited show.

Cyd Charisse was the opposite of Lena Horne. She was a dancer who didn't want to sing. But this was a musical variety show and she would be asked to do a little something with Dean. And of course it had to be live.

Cyd understood the situation and, perfect lady that she is, she put the whole problem in my hands. She confided in me that it wasn't that she didn't want to sing; it was that she thought she wasn't very good at it. Her MGM singing voice was always dubbed and we let Melissa Stafford sing her production numbers on our show. What to do with Dean was a challenge. Remembering Gene Kelly's "I Got Rhythm" number in *An American in Paris*, I literally stole that idea for Cyd. She could speak all the "I got's" and Dean could sing everything else. It even worked with "I Won't Dance."

Cyd was grateful and as we rehearsed the duet, I discovered that her singing voice wasn't bad at all. "What have they been doing to you all these years?" I asked her.

She told me that MGM dubbed in the singing for a lot of people and she got so

Dean and "The Hooker Sisters" (Melissa Stafford, Julie Rinker, and Diana Lee)

used to it she just assumed that's the way it would always be. I told her she could sing or talk or do anything she wanted with Dean and it would be fine. Every time she guested, though, I worked out something that she could take her choice with, singing or talking. She said she couldn't wait to see what I would devise for her next time.

Cyd and her contemporaries in those grand old studio musicals had weeks of rehearsal — plenty of time to get used to whatever demands were handed to them. A TV variety show had to be ready in just a few days. Her dances were nevertheless superb on our show and the talk-sing business with Dean just as satisfying. Besides being very talented, Cyd was and still is a stylish, beautiful lady. Not too many of those in our business.

Choir singer Melissa Stafford's weekly appearances garnered us considerable fan mail. She was a sweet Marilyn Monroe look-alike and Greg gave her some juicy bits in sketches from time to time. This year he suggested that some sort of special presentation was in order.

"People know that face now. Let's give the little lady a break."

Melissa was perfectly happy to continue working in the chorus. I don't think she ever

Dean and
Barbara Eden

considered herself star material and didn't show any signs of pursuing that. I suggested to Greg that she might be more comfortable as part of a trio, with two other charming ladies from Jack Halloran's choir, Diana Lee and Julie Rinker.

In order to attract as much attention as possible, Greg came up with a name for the new trio, "The Hooker Sisters." Naturally the network's Standards and Practices Department took offense at this smarmy name and argued all week long with Greg about it. Finally he backed down and changed it to "Dean's Girls," but not until after calling some press hotshots to witness the arguments and get "The Hooker Sisters" some publicity with both names.

Dean howled when he got word of it just before taping and almost ad-libbed the forbidden title out of sheer naughtiness.

Jimmy Stewart guested again and, since we had conned him into singing and dancing the previous visit, we gave him a double dose of it this time. Jimmy's adventures in varietyland caught the eye of the *Los Angeles Times*, which questioned why on earth an actor of his stature would agree to take part in such shenanigans.

"Some of these other producers ask me to

Dean and Jimmy Stewart

come on their shows," Jimmy answered, "and if I say yes, I'd like to, they go running around asking themselves, 'Migawd, he said yes. Now what're we going to do with him?' Dean's show is done with that live feeling about it. Everybody is dead serious about getting it right, but when the red light goes on the TV cameras on that last day, you never know if it won't blow up. This is exciting, and when it comes off, as it always seems to on Dean's show, if that happens, then it's great."

The season ended with our ratings bouncing back and forth between No. 1 and No. 2.

Everybody, especially Dean and NBC, was very happy. The network was so pleased that they rewrote Dean's contract to read *five years* instead of three. They would have made it more but Dean kept insisting, "Who knows where I'll be in five years? I might be five feet under."

5 The Golddiggers

Greg's production company wasn't going on vacation in the spring of 1968. After Dean's last show of the season he gave everybody a couple of weeks to catch their breath. Then we were back in business for another *Dean Martin Presents* . . . , a catch-all phrase that guaranteed Dean's hold on Thursday night at 10:00.

"This time we need a theme," Greg announced. We kicked around several until one theme made us suddenly state in unison, "That's it!" It was brilliant — we'd do a television show based on the 1930s, but present it as though TV existed then. We'd use only the songs, styles, celebrities, fads, and foibles of that decade.

"We need a title," Greg kept telling us.

I remembered the Warner Brothers musicals of the early 1930s and called out, "The Gold Diggers!" It seemed to say "The Thirties" more than anything else to me. The rest of the staff blurted out another "That's it!" But whenever the typists wrote that title, they seemed to want to run the words together, "The Golddiggers." It was meant to be.

Geoff Clarkson and me rehearsing new Golddiggers

Greg's first choice to star in the show was
Shecky Greene. There were many meetings,
always behind closed doors, in which Greg
used all his persuasive powers to convince
Shecky the summer show would make him a
star.

"I thought I was a star," Shecky kept saying.

"Not on television you ain't!" Greg would
snap back. Shecky hemmed and hawed,
saying he wouldn't be comfortable in sketch
material, his forte up to now being his popular Las Vegas stand-up routines. But Greg
asked him so often he finally relented and
answered yes.

The day before the first rehearsal, he

walked back into Greg's office with the week's script under his arm. "I'm sorry, I can't do it. If you want the truth, I'm scared to death." He stood there quivering, totally submitting to his inner fears of the rigors of weekly television.

"Okay," Greg said quietly, gently taking the script book from under Shecky's arm.

"That's it? No more arguing?"

"No more arguing. Thanks, anyway. We'll keep in touch."

"Can't I even have the script?" Shecky asked shyly. "My name's on it."

Greg shook his head and smiled. Shecky turned and walked out, a tear of regret in his eye.

"I pushed him as far as I could," Greg told us after he left. "And I still say we could have made him very very famous." In the next breath Greg called out to one of his secretaries, "Get Paul Lynde on the phone!"

Paul had been waiting in the wings of Greg's mind ever since the first "I'm not so sure" came out of Shecky's mouth. He'd seen Paul on Broadway in *New Faces* and *Bye Bye Birdie* and liked him.

Knowing the value of TV exposure, Paul came right over. Comedienne Barbara Heller was signed to work as much as possible with Paul, something he regretted from the first day. Joey Heatherton had proven herself as a multitalent on Dean's show, so she and Frank Sinatra Jr., another Greg favorite, were signed for the two musical leads.

But what about that title, *The Golddiggers*? To me and the rest of the staff it meant a bunch of beautiful, starry-eyed chorus girls, not terribly talented but smart enough to snag a rich man.

"Instead of the usual chorus of four girls and four boys, let's find eight sensational-looking broads," Greg said. "No, let's get ten — make it twelve!"

Greg assigned his associate producer, Janet Tighe, to help me do the preliminary search. Janet's credentials were impressive to Greg. She'd worked for Ralph Edwards and *This Is Your Life* for many years. Because of that association, Janet filled Greg's need for special handling of celebrities, which she had done nobly on Dean's show. He also liked Janet's Irish strength and character. And although she had show business smarts, there was none of that movieland gloss he hated. She was the sturdy Midwest counterpart to his Broadway and Hollywood attitude. Just as Greg would often use crude and startling approaches to people and situations, Janet was forever well-mannered, calm, and gracious.

In this role she also had to absorb a lot of Greg's temper, but she knew it was seldom directed at her. Sometimes he needed a punching bag. "I know him better than anyone," she'd say, and her predictions of his actions were usually right on the mark.

Once after yelling at her in a particularly rough way, he sent her a note, as close as he could come to an apology: "I hate to see you with your head down. It's bad for your posture . . . Greg."

Janet would be the one person he could trust to be honest and fair with the young ladies we were about to audition — and still get what he wanted out of them.

"You two look at them first," he instructed Janet and me. "Let me know when you get down to around fifty." He trusted us to that level. Greg was one of the best guys in the business to know a pretty face when he saw one. Attending to the other ingredients was up to Janet and me.

"I'm out of it," Bob Sidney shouted happily. "Whatever you choose, they're just going to stand and sing. Don't expect me to work miracles with choreography."

The requirements were, in descending importance: looks, youth, dancing, singing.

In spite of the fact that our audition ads read . . .

"Incredibly beautiful young ladies
ages 18-22
Must be able to sing and dance"

. . . a cattle call of assorted uglies showed up, most of whom had been pushed into coming by their mothers. Completely ignoring our ad, the age group included girls still in their early teens as well as matronly forty-five-year-olds. When one of the older ones was told that perhaps she might be too mature for the group, she responded with, "Whatever happened to *talent?*"

Janet had the difficult task of saying one of two phrases after each girl auditioned: either "That was very nice, thank you" or simply "That's fine. Will you be able to come back next Thursday?" I sat next to her, making scribbles on each girl's card, using my own rating system for pitch, voice quality, general performance, and appearance. An X in the upper corner meant Janet would politely convey the first phrase. A check in the same corner cued her to make sure we had the young lady's phone number for a return call.

"That was very nice, thank you" was just as difficult for Janet to say as it was for a girl to hear. They knew it meant, "Forget it!"

After some 500 girls had passed by us, we could eliminate certain types right away: for example, the girls who indulged in too many malteds. Or those with the face and figure of Marie Dressler. Or the girl who had been rehearsed in an "audition number" by her coach or mother. Or the singing trio, the sisters who made sure it was understood that they could be broken up if we liked any one of them.

Some girls wore the skimpiest miniskirts in case we were going for all-out sex or maybe just to draw attention away from the fact that their singing voices could curdle your lunch.

A majority of the girls were totally unprepared, not knowing what they would sing or how to present themselves when they came to the rehearsal hall. At one point I made up a list of six songs I thought almost anybody would know, something they could look at and choose from. When a girl seemed at all like a possibility, Janet would make suggestions on makeup and hair styles. However, we still had a room full of silly ladies giggling over their own inadequacies.

"Where do these girls come from?" we asked each other. It was torture on both sides of the table. Auditioning is tough, of course, but as the days went on, it was get-

The original 1968 Golddiggers with Greg

ting harder and harder for us to forgive so much lack of talent.

The tryouts were not a total flop, however. Several very nice things happened. There was a lanky teenager who peeked through the door like a scared mouse, then as soon as her music started became a raging tigress. And the gorgeous girl who couldn't seem to find a key, everything being too high or too low. After helping her search for something singable (we always took more time with the pretty ones, for Greg's sake) she found an area where she could actually hack it. No Peggy Lee, but it was our duty to look for promise.

We finally found twenty gorgeous ladies, none of whom seemed to have any previous professional experience. That was good. We could tailor them our way. We thoroughly expected major problems down the line, but Greg insisted it was "looks 9, everything else 1."

"You and Sidney can make performers out of them," he kept telling me.

We fought for girls with some semblance of talent. "You have to start with something," Sidney kept mumbling. "God, I'll need a magic wand!"

It took us forever to pick the final twelve. Sidney tried getting something passable out of twenty-four left feet. I tried to get as many ladies as possible to sing on key for eight bars. Greg sat and talked with each of them, one by one, for endless periods, to see which ones had the brains and determination to cut it.

"You'll be working your tails off for thirteen weeks! I don't want any married ladies telling me their husbands want them home. I don't want any tears because you think we're pushing you too hard. You're going to work, work, work until you drop!" Greg was playing Warner Baxter in *42nd Street*. Sidney and I knew, of course, that when he said, "work, work, work," it was *we* who'd be doing most of it. Creating routines for Cyd and Juliet and Lena was one thing. Making presentable performers out of this gang of clodhoppers was not going to be easy.

Frank Sinatra Jr. with the 1968 Golddiggers

Since Greg had instilled the wrath of God in them, the twelve finalists worked hard and really wanted to make good. Joey snickered at the lack of competition as far as she was concerned but Paul, with his distinctive disdain, sighed, "Those poor souls need every bit of encouragement they can get!" The girls were so naive to the ways of TV they could hardly contain themselves when we got to their first prerecord. "This little thing is called a microphone," I pointed out. The whole studio was aware that the highly critical sound system could make or break them.

"You'll be able to hear those bad notes nice and loud now," Sidney reminded me sarcastically. The entire cast, the stagehands, Greg, Janet, and Les Brown's band stood around to hear the first playback. It wasn't bad. In fact, it was pretty damned good! The kids had gotten through it.

In that first group, with one of the prettier faces, was actress Cathy Lee Crosby — not strong in the singing department but passable on her feet and sporting a "different" look as far as Greg was concerned. The rest of that first dozen went on to other nonmusical highs and lows, mostly as models.

Chevrolet got wind of our little experiment and signed on as main sponsor of the show. Greg even came up with an idea that the car company loved. We started the show with a 1968 Camaro that the girls completely took apart and put together again as a 1930 Chevy coupe. It took them exactly sixty seconds and a chorus of "Happy Days Are Here Again" to do it. They reversed the process at the end of the show to bring us back to the present day.

Staying with our "TV in the '30s" theme, we opened each show with the Golddiggers as an all-girl band conducted by Joey with

Frank Sinatra Jr., Joey Heatherton, and Paul Lynde on the original Golddiggers show, 1968

Frank Jr. and Paul as "band singers." The sketches dwelled on such '30s subjects as the Lone Ranger and Bonnie and Clyde. The finales were musical salutes to performers of the period such as Al Jolson, Ruth Etting, Maurice Chevalier, Shirley Temple, Eddie Cantor, Rudy Vallee, and Dick Powell.

Paul constantly fought with the writers over every sketch, usually questioning their ability to write at all. He was less critical of Sidney and me and went along with our musical blackouts (taking a popular song and putting it into an entirely different, and funny, situation). "As long as I don't have to work with Heller," he would also tell us. He objected to Barbara because she wouldn't memorize her lines. When she did, she got more laughs than he did. Even if Paul didn't agree, the rest of us thought the show's writing, by Stan Daniels, Tom Tenowich, Ed Scharlach, Peggy Elliot, Al Rogers, and Rich Eustis, was superb, brilliantly and hilariously sticking to the '30s as we'd all planned.

On the show, Paul got along famously with Joey and Frank but had no patience with the Golddiggers. "Imagine putting those untalented girls on nationwide television!" he would exclaim.

To Greg, it didn't matter if the girls couldn't move or sing too well. He'd throw in lots of close-ups and who cared!

While we were still doing our last show that summer, a call came from one of the big hotels in Las Vegas. They'd read all the publicity and asked if the girls could open in their main showroom with the next change of bill.

"Why not?" Greg asked.

"But they don't have an act," I pleaded. I should have known what his answer would be.

"Put one together!" he told me and went hurriedly down the hall to more pressing business.

Dean Martin's Golddiggers opened at the Riviera the following week.

The summer ratings were sensational and we had the top show of the season. Naturally we decided to parade them out again as Dean's summer replacement in 1969. Most of the original girls saw no future for themselves on the road and left before that first Vegas appearance. There was another round of auditions, this time in New York as well as Burbank, with the same sort of gangly, misguided starlets, but enough to fill the vacancies for bookings all over the country and a return to Southern California. Paul Lynde agreed to come back, this time convincing Greg that we could do without Barbara Heller, but the rest of the cast was new.

Bob Hope supervising the Golddiggers' makeup on one of the Christmas tours

Lou Rawls was the male singer and Gail Martin replaced Joey. Stanley Myron Handleman was booked to do monologues but no sketches, Greg once again bowing to Paul. Albert Brooks was hired to do both, but, strangely, didn't make enough of an impression to appear in too many of either. One lanky young man, Tommy Tune, was suggested by Gene Kelly, who more or less discovered him while directing the filming of *Hello Dolly!* Two other Kelly discoveries from the same movie, Joyce Ames and Danny Lockin, were also signed. But it was Tommy who stood out.

Then there was a young lady Greg had been turned onto by some pushy agent. Darlene Carr was young and pretty with a lovely soprano voice and an okay way with the guitar. She was given a solo each week but never appeared elsewhere in the show.

There was fear that a conflict would arise between Darlene and Gail Martin. Greg quickly explained to Gail that he was trying to use some new faces, but Gail was one of those, too. The show was supposed to be her showcase, not Darlene's. An unusual and rather silly compromise was agreed upon. Darlene would not have any orchestrations made for her — instead she would simply sit and sing, accompanying herself on her guitar. The simplicity of that showed her off even more than if she'd been produced. But as usual in these things, everybody got a shot in the ring and they all came off well.

Paul was much happier without Barbara, relying on some of his own personally selected ladies to work with: Carol Cooke, Alice Ghostley, and Allison McKay.

The line of Golddiggers this year was almost all new. There had been many changes while they were on the road: marriage, bit parts in movies, moving on to "better things." Those who stuck it out griped about the rigors of road work and didn't seem to realize that they were becoming better performers every day. They did remember, however, that Greg was in Burbank waiting to put them on national television. So they endured.

Back in beautiful downtown Burbank we had to lead them once again through the discipline of TV, but this bunch seemed superior to the first year's.

"Why can't they just stand and sing?" Sidney still moaned, but Greg wanted a lot of ass-shaking and Sidney did what he could.

After two summer seasons of their own and some guest shots on Dean's show, plus a long string of nightclub dates, the girls made a name for themselves around the country, so much so that Bob Hope asked them to join him on his tour to Vietnam for Christmas. Even though he had Ann-Margret, Linda Bennett, and Miss World, he knew the boys would go berserk over twelve young, gorgeous, gyrating female bodies. Bob asked me to put together something he could do with the girls, and of course the act pretty much followed what they'd done with Dean: stroke and tease, fondle and squeeze. The GIs ate it up. So did Bob. He invited them back five more times.

Johnny Bench was another of Bob's Christmas show guests who found the girls irresistible. In fact, he bet the boys in Les Brown's band that he could go to bed with every one of the girls before the week's tour was over (he only got to two).

Although he endured a shaky reputation as a womanizer, Bob Hope was a father image to the girls. He was there to protect them from all the Johnny Benches and the boys in uniform. He showered them with gifts and made them feel important as people, not just sex symbols.

Whenever they were on *The Dean Martin Show*, Greg not only diplomatically pushed them Dean's way but toward many of the male guests as well. Glenn Ford took a particular liking to Susan Lund. They dated often, and the affair was written up in several of the tabloids. Susie "stayed over" at Glenn's house many times, according to the reporters, but always in a separate room.

The Enquirer made their romance a running soap opera. "Glenn Ford and a Golddigger — 'I adore her and she adores me' " was one of the headlines. "All that she asks of me is that I be myself" was the only quote from Glenn. Being himself, Susie said, was nodding off early.

Glenn pressed Susie for marriage, but she said his grand offer of fame and fortune didn't make up for the fact that she didn't really love him. "And I just couldn't see myself waking up in the morning with *him* next to me!" Susie told me. The affair was mostly platonic. They enjoyed each other's company. He promised her it could stay that way, but she didn't take him up on it.

Dean's favorite Golddigger was Pat Mickey, one of the prettiest girls and perhaps the least talented. He took her out a lot but finally dumped her because she wouldn't go to bed with him. She eventually married one of the Everly Brothers.

Inflation reduced the girls from twelve to ten, and just before flying off to London for a summer show, Greg sliced it further to just eight. Sir Lew Grade had shown interest in getting the Golddiggers across the Atlantic and offered an up-front price Greg couldn't refuse. Janet and I, plus choreographer Jonathan Lucas, flew over several weeks ahead of the girls to set up housing and get the show ready. Marty Feldman was Sir Lew's choice to star, but Greg felt an American comic was necessary and brought along Charles Nelson Reilly to costar. We all entered the country as "creative advisors" so the English performers' unions wouldn't ban us altogether.

Jonathan and I found time to audition some English girls for possible additions or replacements. We took over London's famous Talk Of The Town nightclub one day and waited for what we thought would be hordes of girls wanting to join the group and possibly "go to America!" Only about fifty showed up, the lure of California not as strong as we presumed. We found two young ladies, Pauline Antony and Wanda Bailey, who were very talented and eager to do our TV show in their own country. But when we brought up the subject of joining the group in the States they weren't too sure.

As the tapings went on, though, and they

Glenn Ford and the Golddiggers

got to know the American girls and hear their stories of glamour and excitement among the rich and famous of Hollywood, they later agreed to come back with us.

The rest of the Golddiggers — Jackie Chidsey, Paula Cinko, Susie Lund, Rosetta Cox, Michelle Della Fave, Tara Leigh, Pat Mickey, and Micki Malone — were the most talented we'd had so far. Greg even said he'd give each of them a solo spot, one on each show.

Such squeals and screams greeted the announcement that you'd have thought we would be creating automatic stars. Well, after all, they'd just been a glorified bunch of chorus girls up to this point.

Unfortunately, Greg announced this just

The 1970 Golddiggers

a day before our first taping, so I had to pick someone to do a fast solo with very little rehearsal. I hoped she'd be able to pull it off and Greg wouldn't decide to chuck the idea.

My choice was a good one. Paula Cinko was a big girl with a big voice and a sweet personality. We worked on "When I'm Not Near the Boy I Love" as much as we could between other songs and sketches, and when it came time for her big chance she hit all the right notes. Greg was pleased, mumbling that at last we had more than just a bunch of pretty faces.

221

Elke Sommer, who had guested on Dean's show and was in the same studios in Elstree doing a Burt Bacharach special, would drop in to see what we were doing. As sexy and provocative as our girls were around the studio halls, Elke was something else. She loved wearing see-through blouses and attracted a great deal of attention whenever she bounced by.

One evening after rehearsals we all went to a little Italian restaurant in Soho. It was a hot night and the sign outside said "Air Conditioned." As often happens in London when it's sweltering, that promise doesn't always work out. But Elke and our gang decided it was worth the sweat. The Gold-diggers who joined us were, as per Greg's orders, the picture of decorum. When Elke, on the other hand, waltzed in wearing a revealing dress, the waiters broke out in a loud round of bottle-hitting. (The ceiling of the place was decorated with empty wine bottles, and the staff would delight in hitting them with spoons whenever they wanted to attract attention.)

In fact, whenever they passed our table there was another round of bottle-smacking. When we left, the waiters refused to take a tip. They'd been gratuited to a tee.

Outside in the street, Jonathan Lucas said

he was too hot and proceeded to take off his shirt. Not to be outdone, Elke went topless, too. The other girls ran off to the Underground for fear Greg might be watching, but the rest of us made quite a scene in Soho. Elke's husband, writer Joe Hyams, found his Fiat and the four of us piled in, Elke standing bare chested out of the sunroof, waving her blouse to an admiring crowd of cheering young men. A traffic jam ensued and it seemed for a moment that the entire population of Soho might assault her. Joe managed to gun the Fiat just in time. Ah, *la dolce vita!*

The biggest boost we got while in London was Greg's decision to take a chance with Tommy Tune as a "new face" TV audiences would be interested in. He knew Tommy was a talented dancer, but he also figured his unusual, six-foot-six, height and boyish manner would pay off on-camera. He was right. Tommy not only got a classy solo spot singing and dancing on each show but worked in sketches with Marty and Charles and assisted Jonathan on choreography, too.

His imaginative staging gave the show a definite plus, and Sir Grade provided us with nothing but the best sets, costumes, makeup, and hairstyles. Tommy's socko

The
Golddiggers
in London

numbers were conceived, he said, while getting high the night before each production meeting. "All sorts of Ziegfeld numbers just sort of flash into my head," he said dreamily. Whatever the cause, there seemed to be no end to Tommy's originality and brilliance.

Charles and Marty didn't get along at all and both of them went to Greg to ask not to work together. Charles couldn't tolerate Marty's frantic comedy, yet it was that insanity that allowed Marty to come off as something new and different.

Marty would go absolutely berserk as soon as the cameras rolled, especially if there was a studio audience to react to. In a sketch that called for him to remove all of Susie Lund's makeup (deliberately heavy)

Tommy Tune

as well as her dress (a breakaway that would reveal her in scanties), he almost ripped her real eyelashes off in his wild abandon. Each take was followed by poor Susie having to go back into the makeup room to have everything, including the heavy lashes, put back on. After hearing the wardrobe department warn him that he would have to pull Susie's dress at a certain spot (a brooch at the top of her cleavage), he wickedly pulled it everywhere else and eventually tore it all off, leaving Susie stark-naked in front of the studio audience. As this was a British audience, they loved it. Susie was in tears and refused to do it again.

That wasn't the limit of Marty's hazardous behavior. Poor Charlie was a patient in a sketch in which a nurse tried to talk him into going ahead with an operation by a notorious drunk doctor (Marty). "He's only had one quart of gin all day — this may be his last chance," the nurse confides.

Charlie decided to give him that chance. When Dr. Feldman exploded on the scene, knocking over tables of medicines and surgical equipment, he began the operation by asking for a jar of knives and sutures. In typical Marty madness, he flung the full jar over Charlie's body. All sorts of dangerous razor-sharp utensils came crashing down on Charlie. That was it. No more skits with Marty.

The show was a summer hit in the USA, but due to contractual problems all of Marty's sketches were removed from *The Golddiggers in London* shows aired in England. The British audiences didn't take to Charles and felt the girls weren't all that talented. They liked Tommy, but without Marty, the show was not a hit there.

On returning to Burbank for Dean's fall start-up, the girls were divided into two groups, those to be regulars on the Martin show and those thrown out on the road to make money. Naturally the road-bound Golddiggers were less than thrilled about their fate, but Greg promised them he would switch groups somewhere down the line, which he did.

Having two groups meant more searches. Janet and I hit the road again, this time with choreographer Ed Kerrigan, and set up auditions in Dallas, Chicago, Boston, and New York.

One of the highlights of the off-Broadway season that year was a little satire called *Dames at Sea*. We were told it was a spoof of the same Warner Brothers musicals of the '30s that we had used as inspiration on our Golddiggers shows, so we headed down to the Village to take a look. The star was a vo-

Tommy Tune, Charles Nelson Reilly, and two of the Golddiggers in London

luptuous young lady named Bernadette Peters. We liked her a lot and went backstage to see if she might be interested in chucking all this little-theater nonsense and coming to California to be a Golddigger.

Greg used all his suave powers to convince her that this would be the turning point in her career. We took her to dinner at a fancy Chinese restaurant on Seventh Avenue and told her what tastes of celebrity lay ahead. She was impressed with the pitch and said she'd think it over.

We were sure we had her locked up and that she would be a welcome addition to the group of eight now settled in on Dean's show. She even said that she was leaving her show anyway and loved the idea of going back to California. The next day we didn't hear from her, so Janet called the number she had given us and Bernadette said that she had decided she'd rather try it on her own. Greg wasn't through yet. He asked her to dinner again, but she turned the invitation down. Bernadette, of course, went on to be the darling of Broadway, starring in many musical shows and winning two Tony awards so far.

Instead of having the Golddiggers as summer fill-ins for Dean again, Chevrolet asked us to tape thirty half-hour Golddiggers shows for syndication. CBS picked them up immediately for their major outlets and put them in the 7:30–8:00 p.m. pre-prime time slot. Charles Nelson Reilly would once again be the regular comic.

Chevrolet's enthusiasm for sponsoring the girls was dimmed somewhat by its insistence that each Golddigger sign a multipage morals clause that forbade them from indulging in anything naughtier than a hot fudge sundae. Their private lives would be closely watched so that no scandal would sully Chevrolet's hallowed name. A little smoking and drinking was okay, but everything else was out. The poor girls were petrified at first, but Greg had already laid down similar laws. From the beginning, for instance, he forbade them from leaving their homes or hotel rooms without full makeup. He pointed out that they were public property now and their fans would expect them to look glamorous at all times.

The girls almost *had* to agree with Chevrolet's rules. Thirty shows are thirty shows. And thirty paychecks. We taped all thirty in five weeks at the Hollywood Palace in assembly-line style. I had *one day* to teach the girls *thirty numbers* — an opening with that week's guest, a group production number, a solo from one of the girls, a "pretty spot" in which they could just sit and sing and look captivating, and a finale with the week's guest for each show. It took two days to stage everything, one day to block and tape the sketches and another to do the same with the thirty songs.

Taping days resembled a Marx Brothers movie. The girls arrived at 5 a.m. for makeup and were called on-camera two hours later for the first song. As soon as that was taped, they'd rush back to the wardrobe department to make a quick change for the next number. While they were doing that I'd explain to the scenic designer what the next song was about and he'd scrounge around the limited set department for columns and steps, then get the stagehands to bring them out just as the girls appeared. That routine was repeated 30 times during the day until all the numbers for six shows were on tape.

Fess Parker, the tall, soft-spoken western star, guested one week and surprised us all with his pleasant baritone. He very nicely handled a musical staging of "Small World" from *Gypsy*, in which a computer date error teamed him with Susie Lund, the smallest girl in the group.

With one guest per show, almost always a man, the shows looked much like a small-town version of Dean's show. Ernie Borgnine, Hugh O'Brien, Mike Conners, Don Meredith, Doug McClure, Martin Milner, and George Maharis were some of the hosts who came off better than most.

We were lucky with George Maharis. He had not only sung in summer stock but had just made a couple of best-selling albums. The girls thought he was the sexiest of all their guests, but he did nothing about it after he left the studio and, of course, the girls had been warned by Chevrolet not to mess around.

The *Golddiggers* proved another winner in syndication, the girls themselves getting the best reviews. Critics gave all the flack to the writing. "Old hat and predictable," they wrote, "proving only that the writers have good memories." (Many of the sketches were rewritten from *The Dean Martin Show*.) Chevrolet was happy, however, and that's what mattered most.

A year later Chevy looked over the ratings of that first syndicated series and asked for another thirty shows. Once again the girls surrounded a male guest on each show, gave him the old kiss-and-punch-and-grope routine, which sometimes bordered on outright silliness.

Carol Burnett was still doing her weekly variety show over at CBS and used the Golddiggers' first album to conduct lunchtime aerobics sessions with her cast and crew. The girls' show also caught her eye for parody and she did a devastating takeoff on it. She called them "The Goldgigglers" and attacked all the obvious banalities, particularly zeroing in on the girls' inability to read cue cards properly. She had a good old time putting herself and all the other ladies on her show in outrageously teased hairdos, mugging sinfully into the camera whenever they felt they might have a close-up. As on *The Golddiggers*, she had a macho guest of the week (William Conrad) whom the girls fondled and fussed with as he was trying to read the Gettysburg Address. Everybody on our show loved the sketch and felt honored that Carol would see fit to acknowledge us so hilariously.

The last regular *Golddiggers* season was in 1973. They went on to be grandly featured in several of Dean's specials, some spectaculars with the likes of Gene Kelly and Tennessee Ernie Ford, an Emmy show, and of course Bob Hope's Christmas trips to Vietnam.

Over 75 girls were Golddiggers at one time or another. We were constantly re-

Me, last in line, as an extra with John Forsythe on one of the Golddiggers syndicated shows

placing somebody who found greener pastures elsewhere. Although the group started as an even dozen, they ended up as a quartet opening for Dean at his monthly dates in Las Vegas. Those last four girls were definitely among the best: Peggy Gohl, Marie Halton, Maria Alberici, and Linda Snook. In 1991 the group sang their last sweet notes, tearfully saying goodbye to Dean and the rest of the country at Bally's.

Most viewers considered the Golddiggers merely a step or two above the Laker Girls. But our beauties had something more. They were the envy of every young American girl and the dream of every hot-blooded American male of any age.

They were a lot of things to a lot of people. (Johnny Carson referred to them as "Garrison's Gorillas.") Their talents were modest. They could sing a little, dance a little, pose a little, and look absolutely stunning on-camera. They also represented the epitome of male chauvinism to women's lib groups, one of which picketed our offices on the corner of Riverside Drive and Hollywood Way in Burbank. Greg thought it was great. He called NBC and got their cameramen over to record the scene for the 6 o'clock news.

"Hey, you can't beat that kind of publicity!" he reminded all of us.

It was a great run. Oddly, most of the girls don't give credit for their success to Greg or to me. Like most of the world, they still think it was Dean Martin himself who made them what they were. Well, let's face it, they were featured on *his* show and he showed off their attributes better than anyone else — to which Dean would have said, "I did what?"

6 The Fourth Year – (1968–1969)

Reaching the Peak

"Instead of more guests, let's work the ones we have more. Give me lots of sketches."

That was the 1968 order from Greg for the new season. "And let's mix it up. One week we'll start with a song, the next week with a sketch. Let's keep it all loose and unpredictable. I don't want people saying, 'At 10:15 I know I can go into the kitchen because that's when the girl singer comes on.' "

The first show of our fourth season had another stellar lineup: Buddy Ebsen, Shecky Greene, Barbara Heller, Lena Horne, and Zero Mostel (or Zero Muscatel, as Dean kept calling him). Zero was from Broadway and that was sometimes a problem. It usually got a turned-up nose from Dean (incredibly, he *never* saw a Broadway show!).

But Zero's zaniness completely captivated Dean, especially his ability to more than match Dean's ad-libs. Their moments to-

Zero Mostel and Dean

gether, albeit short, were gems. Dean's only objection was that Zero sweated so much. Greg didn't like the way that looked on-camera so we'd have to interrupt the flow of the show from time to time to mop up. Dean didn't really find the perspiration all that offensive; he just didn't like the fact that we had to stop.

"He's sweating on my time," he chuckled.

Dean instinctively knew that Greg would never allow such lapses of taste on his show. The two of them were of one mind about that. Yet once, something that Greg and I both found quite innocent prompted Dean to "suggest" a change. One of the writers had come up with the idea of doing a garbage medley. Things like: "Tip Toe Through the Garbage," "I Love Garbage in the Springtime."

Dean watched me do it on the monitor in

his dressing room and we were interrupted by a messenger asking Greg and me to please come and see him.

"I don't really want to be sitting at home having dinner and hear a bunch of songs about garbage," our suddenly aware star sternly told us. "That's more than I can stomach," he added with a bit of a grin.

We were so impressed with the fact that Dean actually cared that much that we rushed out and changed every "garbage" to the word "rubbish." It didn't make the point as well, but it made us all feel better.

Lena Horne guested again that year and sang her $10,000 song (her asking price from the previous year) for $7,500, but without a dance number. No more was said about her earlier demands. I put together a long medley of current pop tunes, several of which Dean had actually learned for an album. He usually balked at anything so new but the quality of some ballads that year had caught his fancy: "By the Time I Get to Phoenix," "Little Green Apples," and "Honey." Lena had to learn the borderline country hits, which she did quickly and brilliantly. She was always able to turn a good song into a great one. At the end of the medley, Dean held Lena's hand, a small gesture, but in those days the networks not

Dean and Lena Horne

only frowned upon whites and African-Americans touching, but banned it out-right. Greg, outraged by this ridiculous rul-ing, zoomed in on the hand-holding for a close-up. The NBC censors didn't like it, but they let it go because these particular stars could "do no wrong."

Ageless hoofer Buddy Ebsen did one of his tap routines, part of which was one called "eccentric dancing," a snakelike movement of the body, almost in contor-tion. I knew Dean had been exposed to this sort of thing by his uncle, Leonard Barr. Whenever Uncle Leonard was on our show and wound up his monologue with a chorus of "Crazy Rhythm," Dean joined him on the last eight bars. "Same old routine he's done for forty years," Dean would mumble under his breath.

Bob Sidney and I devised a way to get

Dean to follow Buddy through a routine, by watching and copying him. It was the kind of thing that couldn't be rehearsed. Thanks to Uncle Leonard, Dean added some eccentric movements of his own.

Zero Mostel aside, Greg liked the sweat and heavy breathing that followed most dancers' big numbers on the show. He didn't like to stop tape. But Buddy was getting on in years, and we did stop to let him catch up.

Buddy had worked with Sidney's assistant, Wisa D'Orso, on a previous show and asked for her again this time. Wisa was more than a choreographer's helper and Buddy knew that. He considered her one of his best partners ever. Dean was so impressed with her that he asked Greg why he didn't give her a solo spot on the show. To our amazement, he did. Sidney and I fashioned a big production for her and the rest of the singers and dancers. Like all our guest stars, she also did a song with Dean, who always seemed to thoroughly enjoy giving a talented unknown a chance.

After three years of Dean jumping on Ken Lane's piano every week, we got the special effects department to make a breakaway piano of balsa wood that would completely

Dean
and
Buddy
Ebsen

collapse when Dean hit it. Naturally, it couldn't be tested in advance. Usually when we had anything remotely dangerous for Dean, Greg would try it out himself first. But this time Dean was let in on the gag. He was briefed about what might happen and assured by the makers that it was perfectly safe.

"I'd do it first," Greg told him, "but we've only got one of 'em."

"Hey, don't worry, I'm ready," Dean answered. "If I break my ass, Lee can take over."

Dean jumped as usual. The piano gave way and the audience gasped. But when they saw that Dean was all right and actually enjoying it, they laughed and applauded for three minutes. Greg just let Dean lie there, amid the broken-up Steinway pieces. He milked as much out of it as possible.

"My tutu hurts," was Dean's only comment as he was led back to his dressing room for a tux change.

Greg also decided that Sunday tapings were out, that we all deserved that day off, so we shifted the tapings to Saturdays. Dean couldn't have cared less. He fussed a bit about the fact that he'd be stuck with college football in his dressing room instead of the more exciting NFL games on Sunday, which he always had money on.

"If the Rams win, we'll have a good show," Greg had maintained. "If they lose, we'll have to get him going again." Maybe it was just as well we didn't tape Sundays anymore.

It was also the year we inaugurated another "department," this one involving the girls again, something called "Musical Questions." Each girl would sing a line from a song and Dean would answer with a lyric from another song. Since this had to happen quickly I worried a lot, waiting to see if Dean would find the right starting notes. If he didn't, he'd somehow fake it and we'd sail along.

Barbara: *Give me a little kiss, will ya, huh?*
Dean: *All or nothing at all.*

Roberta: *I'm in the mood for love . . .*
Dean: *I'd like to get you on a slow boat to China.*
Ellen: *I am sixteen going on seventeen . . .*
Dean: *This nearly was mine.*
Jeri: *I'm just a girl who can't say No . . .*
Dean: *You'd be so easy to love.*
Kate: *Cigarettes and whiskey and wild wild women . . .*
Dean: *These are a few of my favorite things!*

On one show Melissa Stafford was the third girl down the steps and sang the wrong line of "I Get a Kick Out of You." Instead of *"I get no kick from champagne,"* it came out *"Mere alcohol doesn't thrill me at all."* She panicked because she knew Dean's answer wouldn't come out right.

Dean jumping on Ken Lane's piano

Dean was jubilant: "She made a mistake! I didn't make it, she did!"

Melissa began biting her nails.

"Don't do that," Dean ad-libbed. "You know what happened to Venus de Milo! Well, you're off." He motioned for the next girl to come down, then changed his mind. He brought Melissa back.

"No, honey, we'll do it all again from the top, just for you." He stopped the band, which had been roaring along as per weekly instructions from Greg never to stop unless he gave the word. "Keep it going," Greg yelled into the crew's earphones. It was one of those scenes Dean could handle so charmingly, the kind of thing that made the show different.

Dean cuddled Melissa in his arms.

"It's all right, honey, it'll just come out of your salary."

The girls sauntered up the stairs to start all over again.

Dean kept it light. "We all make mistakes. Remember Hitler?" The music began again. The first two girls repeated their bits. Dean got to one of his answers and he confided to his viewers, "I put more into it that time." When Melissa got out the correct line, *"I get no kick from champagne,"* Dean finally gave his answer, *"Two different worlds, we live in two different worlds. "* The audience loved it.

Dean's opening monologues and various introductions were just as spontaneous. Because of his ad-libbing, show timings went out the window. But Dean never bothered with little details like that. Mae West must have been a fan — she liked a man who took his time.

Musical Questions

In the Ken Lane spots at the piano, Dean liked to throw in one-liners that weren't on the cards, something writer Harry Crane might have given him as he was walking out to the stage, like "In the famous words of Sophia Loren who said to Twiggy, 'How's every little thing?' "

Orson Welles guested several more times that season, each time getting more used to

Dean's style and enjoying himself a lot. On one of our drives to and from the Beverly Hills Hotel, I indulged in more questioning about the reactions of the Hearsts to *Citizen Kane*.

"Is it true that they sent you to South America to get out of town, as it were?" I asked.

"Hearst was applying lots of pressure. He even tried to get the film destroyed. I was getting threats on my life everywhere I went. Nelson Rockefeller suggested RKO find me something to do in Brazil, so I went down and made a documentary on current lifestyles there. They never released it, but it kept me away from Hollywood for almost a year."

I asked him if Brazilian music was a part of that picture. "Yes," Orson continued, "I actually had a lot of fun with something called the bossa nova. So many new and different musical instruments were involved. I loved it, but RKO didn't understand it at all. It didn't reach this country until twenty years later."

I visualized a production number with the great Orson Welles doing a bossa nova with all the singers and dancers. Dare I ask him? I dared.

"Listen," he said, "I think it's terrific that you guys get stiff old dramatic cronies like me to sing a little and do sketches, but I'm afraid that doing a song and dance number is out of my league."

"You could just stand and play the bongos."

"I used to play the violin — wasn't bad, either. I was very young. Mother introduced me to Maurice Ravel once." It sounded like he was changing the subject.

We didn't discuss it anymore that time, but when I picked him up for a later show, he threw me a curve. "Well, what about the bongo number?"

"I didn't think you were interested."

"Nonsense. It might be fun. Let's give it a try. We can always chuck it if it doesn't work out."

He said he remembered a song called "So Danco Samba." I constructed a little "Orson Sings Eight Bars / Orson Plays Bongos / Girls Dance / Orson Sings Last Eight Bars" routine that was about as simple as possible.

"It's short. That'll help," he said. He also insisted on a sort of disclaimer which he wrote himself: "Dean sent me out here because I foolishly admitted that I'd been to the carnival in Rio. He didn't realize that they didn't ask me down there to sing. This is actually my first musical solo in any me-

246

Orson Welles getting his own production number on Dean's show

dium — another great last for television!"

He wanted me to stand close by to remind him when to come in, something I was used to doing with every guest as well as with Dean. But he didn't need me. As usual, Orson knew what he was doing.

He autographed a picture for me with the caption: "To Lee, who put me where I am today in the World of Music."

Although Orson didn't quite agree, I considered the experiment successful enough to give it another try the next time he showed up. He went along with it, but insisted he'd have to do more than just play the bongos. I scrounged up half a dozen other Brazilian percussion instruments for him to play and set him up in a big hammock with the girls samba-ing around him to the tune of "Calypso Pete."

Unfortunately the show's taping went slower than usual that day and Orson's number was rescheduled to be the last item, to allow Dean and the other guests to get out of the studio earlier.

Orson wasn't happy with this change and went to his dressing room to sulk. And drink. When it was finally time to tape his number he was pretty well soused. Onstage, as we waited for lighting and other delays, he proceeded to break, one by one, every instrument we gave him. The prop department had several extras ready, but he broke those, too. Orson decided he'd had enough anyway. He somehow got himself out of the hammock without falling and swayed back to his dressing room.

We didn't ask him to come back onstage. The girls were released, the taping was wrapped, and the number abandoned forever.

Greg was constantly trying to make the show different from other variety series and he was getting bored with the same old guests who popped up on all the other shows as well as ours. He started searching for unusual bookings.

"Get me some sexy actresses!" he told

Elke Sommer

Henry Frankel, the show's booker. Elke Sommer was the first of the lot and luckily for me, she had a great deal of musical ability. She also knew her way around a comedy line and certainly fell into Greg's category of sexy.

Her first appearance in our rehearsal hall almost caused a riot. She arrived in one of her see-through blouses and half the studio employees came by to watch. We closed the doors for privacy, but a gang of voyeurs still waited outside, just in case someone would enter or leave and they could get another peek. Sidney and I decided to put her in the most provocative situation we could think of.

Elke used her husky Lauren Bacall voice

to slither and squirm through a slow, sexy version of "Thank Heaven for Little Boys," the boy dancers rubbing next to her, she caressing them from head to toe. We had to clean it up a bit for the censors, who smelled something naughty in the making and hung around just to make sure we didn't go too far . . . or so they said.

Strangely enough, even though we did get away with a lot, reaction from the viewers had nothing to do with sex. They were concerned about a bear rug we'd used for Elke to maneuver on. Mail from animal lovers across America objected to seeing "a dead animal on-camera." To compound matters with NBC's Standards and Practices Department, a medley by Dean and Elke that followed her number caused just as much a commotion. It was set up as a scene in which we were to imagine that the two of them were in a nudist camp and totally naked.

Greg asked me to put in as much raunch as I could, to overdo it so the censors would have lots to cut, but what they left in was still choice:

Dean: *Jeepers creepers*
Where'd you get those peepers?
Elke: *I got plenty of nothin' and nothin's plenty for me!*

Greg had passed Angie Dickinson in the halls of NBC and pleaded with her for days to do our show. "I'm not a musical performer," she kept saying. "I'd be petrified."

"Lee'll make it easy for you," he insisted, "and Dean and I really think you'd be

Dean and
Angie Dickinson

great." Greg usually included Dean in his pep talks, even though our star knew nothing of it.

Angie had just done a movie with Dean, liked him, said she thought our show was wonderful, but she still couldn't be talked into doing it. She began to avoid Greg every time she saw him coming. She avoided me, too, knowing I'd pick up where Greg left off.

One day Greg sneaked up on her again.

"Give it a try. If you don't like it, we won't

use it. That's a promise."

After she finally gave in, we had to think of something she'd be comfortable with. Angie was married to Burt Bacharach at the time and I kept pressing for one of his songs.

"My God, he'd kill me. He knows what kind of singer I am." Burt was in Boston with the pre-Broadway tour of his show, *Promises, Promises,* and the small talk in the rehearsal hall got around to how the show was going.

"Oh, it's great," Angie said, "Burt called last night and sang me a cute little song they just put into the show."

Ah-hah! "Why don't we do that?" I asked hopefully.

"I couldn't. He just sang it for the first time."

"But you said it was a cute little tune. It can't be that difficult to learn."

Angie Dickinson is surely one of the nicest ladies in the business. We wanted to do the very best for her and this seemed like a natural, so we pressed a little.

She began to weaken. "Oh, you people are too much. I suppose I could call Burt and ask him about it." I offered to speak to him, that if he agreed to let Angie do it, I'd take down the words and music on the phone.

Pat Boone, Angie Dickinson, and Dean

He halfheartedly gave his approval and proceeded to sing the song and made sure I had all the chords right. It didn't take long to teach it to Angie, and Sidney staged it nicely with four boys dancing around her. She was the first performer on TV to sing Burt's big hit "I'll Never Fall in Love Again."

Many years later I asked her about it. "Burt was so upset," she confided. "He absolutely hated it." She said she just couldn't watch the show when it aired. She assumed it was a total disaster. Although she's divorced from Burt now, she was afraid to look at a cassette copy I offered her. "It's wonderful," I told her, but she didn't want to see it. Every time I ran into her I said she would be pleased at how well she'd done. She eventually agreed to accept the tape and take a look. She phoned me the next day. "I had to call that son-of-bitch and tell him he made a nervous wreck out of me all these years. It was damned good! Thank you for making an unmusical person very happy."

When we did the show, Angie said she was happier doing the sketches with Dean and Pat Boone. The three of them couldn't have had a better time, except that Greg might have tried too hard to take away some of Pat's "milk and cookies" image. "Let's see if we can sex him up for at least an hour on Thursday night," Greg tittered.

Addressing Greg's wishes, I picked a song from *The Night They Raided Minsky's* called "Take Ten Terrific Girls (And Only Nine Costumes)." Pat didn't see the humor at all when I suggested he could be the production singer in a burlesque show setting. He didn't mind the role so much, but what about all those scantily clad girls slithering around him? Even though this was *The Dean Martin Show* and that sort of thing was par for the course, Pat felt very uncomfortable and begged us to change the whole concept. He simply felt that it just wasn't *him*. But that's exactly what Greg liked about it. We compromised with Pat's suggestion that he could be the stage manager, commenting on all that was going on but staying safely over on the side.

Greg couldn't resist having Sidney secretly stage an alternate ending with all the scantily clad girls pouncing on Pat, much to his surprise and embarrassment. He seemed to enjoy it. But he never came back on the show.

Another bombshell from Europe was the Italian film star Gina Lollobrigida. Somewhere along her Via Veneto way, she had thought she would become an opera star. I'd read about that, and I was concerned about how that could be used on our show. I was prepared to find a suitable Puccini aria when she informed me her first day that she now had developed a sexy lower voice and would be happy to show it off to us.

I chose "C'est Magnifique" to fill that bill

Gina Lollobrigida and Dean

255

and she was everything she claimed — sexy and beyond.

The writers took advantage of her lack of knowing everything there was to know about Americans by having Dean explain baseball:

"It's our national pastime," Dean told her.

"In Italy, the national pastime is love. And it's much better. You don't even need a uniform."

Gina made frequent visits to Dean's dressing room. But long after the show had finished taping, thinking everybody was now out of the studio, I was surprised to hear crying as I passed by her dressing room. I knocked on the door, never thinking Gina would still be there, but she was sobbing her heart out over something she said had happened "during the show."

"What was it?" I asked. "Maybe I can help."

"It's nothing." But then she broke down and told me that nobody loved her, most of all Dean. He had run off after the show and left her there and she didn't have anyone to go back to the hotel with.

I offered to take her, but she had called her manager and told him that "certain plans" had been changed, and he was due any minute. I felt so sorry for her, feeling as she did that none of us appreciated her.

I had figured that maybe we hadn't thrown enough bouquets at her after her number and her solid baseball routine with Dean, so I proceeded to praise all of that over and over.

She was grateful, and although she had told me earlier that she would "never be on this show again," when her manager showed up, she insisted he book a repeat show as soon as possible. And she did come back, often. We treated her better the next time. As for Dean, he was very nice to her, but he still ran away as soon as the taping finished.

Victor Borge's sophisticated musical routines seemed ideal for our show, but what to do with him and Dean? A section of his concerts had to do with vocal punctuation marks. Perhaps we could get him to explain that to Dean.

"You can't *see* a comma or a period when you speak, so my system allows perfect meaning to every sentence by making certain sounds when those punctuations appear on the cue cards," explained Victor. He happily agreed to not only let us use the idea but to transform it into a medley of songs for Dean.

Dean and Victor Borge

To Irving Berlin's "Remember":

Dean: *Remember the night,*
Victor: *(comma sound)*
Dean: *The night you said,*
Victor: *(comma sound, quotation mark sound)*
Dean: *"I love you",* —
Victor: *(quotation mark sound, comma sound, dash sound)*
Dean: *Remember?*
Victor: *(question mark sound)*

And so it went, through several other songs. Victor was asked back often and given both solo pieces and long sessions with Dean.

The search was on for sexy actresses like Elke and strong male actors like Jimmy Stewart. The late Milburn Stone of *Gun-*

Dean breaking up at Milburn Stone's spoonerisms

smoke was a particularly good choice. He could sing, of course, and also brought along some fancy pieces of his own material, which we made use of with Dean.

One of the most hilarious was Milburn's "mixed-up words," or "spoonerisms."

Janet Tighe ran to the dictionary and found this definition:

"A slip of the tongue whereby initial or other sounds of words are accidentally transposed, as in 'our queer old dean' for 'our dear old queen' — named after Rev. W. A. Spooner of Oxford, noted for such slips."

We decided not to use that *Collegiate Dictionary* example but instead had Milburn tell Dean the story of "Prinderella and the Cince."

"Oh, I'm just fairy about crazy tales," Dean was quick to ad-lib.

Milburn took over. "You see, Dean,

Prinderella did all the worty dirk while her sisty uglers sat around on their fig bannies."

Dean kept up with him. "That's a shirty dame."

"Then there appeared a magnificent colden goach made of a pipe rellow yumpkin. All of a sudden there was a flinding blash of light."

"NO SHIT!"

The house fell in. Milburn couldn't go on.

"I'm sorry," Dean kept saying through his tears, "I couldn't help myself."

Three minutes later they pulled themselves together and went on with the story. "No Shit!" was simply bleeped out and the whole bit left in the final show. It wasn't hard for the audience to figure out exactly what words Dean had yelled.

Milburn was a joy to work with, both musically and with Dean in sketches. He was truly one of the all-time nice people around. His death in 1980 robbed many of us of a good friend.

We continued to look for dramatic performers for possible guest shots. Michael Landon was still starring on *Bonanza* at the time and we'd been told he could handle a musical number, that he'd shaken a leg or two on *Hullabaloo* not too long ago. When

Dean and
Michael
Landon

he arrived for our first rehearsal he announced he was leaving everything to me, that he had nothing whatever in mind, and he was our slave for the week.

For some reason his appearance and manner reminded me of the part of Gaylord Ravenal in *Show Boat*, the handsome riverboat gambler who knew his way around women as well as a song. I suggested a minstrel medley of sorts, with Michael singing the well-known hits of that era such as "Rockabye Your Baby," "Swanee," and "Waitin' for the Robert E. Lee." The singers and dancers would set it up behind him with tambourines and drums.

It sounded good to Michael, but he said he'd need all the help he could get in the vocal department. I gave him that, plus the opportunity to spend as much time as he wanted perfecting the prerecord. His singing voice was pleasant enough and he knew when he went off-key, so we did it in pieces, stopping here and there to do a better take, then letting the soundman put the best parts together.

The number came off extremely well and I added similar songs for a Dixie medley with Dean. This time, though, Michael would have to sing it live. He was shaky about it, but I put a little bug in Dean's ear about Michael's nervousness and Dean went out of his way to help him along. It didn't take more than a word or two from me for Dean to sense that a nonsinger might need a little more attention from him.

Michael was glad when it was over but, having that success under his belt, asked to be invited back, which we did on several occasions.

"Believe it or not," Michael confided to me, "I sang on the very first *Bonanza* and they said the show would never be a hit if I did that again. Look what happened to us!"

Our last show of the season headlined Jimmy Stewart with Raquel Welch, the latest in sex symbol guests. Bob Sidney knew Raquel from the *Hollywood Palace*, where she got her start as a buxom card girl who introduced each guest. She had graduated to movie stardom, of course, but Sidney wasn't sure what she could do musically. Not much, we were told by our peers.

To insure that there'd be no problem, Sidney and I devised a "star-proof" sexy version of "Show Me" from *My Fair Lady*, so simple that we could take any girl off the street and teach it to her in half an hour.

The boy dancers were rehearsed in advance to help her move this way and that, so easy that she wouldn't even have to think.

The minute she arrived at the rehearsal hall, she insisted on seeing her dressing room before anything else. I tried to explain to her that this was Monday and dressing rooms were not assigned until the day we went into the studio. "Other shows have to use the dressing rooms," I told her. She didn't care about that and she demanded to see whatever room was going to be hers or she wouldn't rehearse. We were off to a great start.

I finally called and asked someone to show her a room, any room, knowing full well it might not be hers at the end of the week. The moment we entered it, she objected to the fact that it had fluorescent lighting. She refused to go back and listen to "Show Me" until somebody would guar-

263

antee to put in regular lighting. I called one of our secretaries, who promised to take care of it, and we finally got down to our Raquel Welch production number.

She didn't mind the song, she said, but hated the background I'd come up with and liked Sidney's staging even less. She said she didn't want to discuss it anymore, left the rehearsal hall in a huff, and drove herself to Greg's office to tell him how unhappy she was and that Sidney and I should be fired for incompetence.

"Sure, whatever you want," Greg said. He more or less ignored her. This was our final show of the season and he didn't want to spend time trying to find a last-minute replacement for Raquel.

Greg chuckled as she left for the day, making sure that Sidney and I should pay no attention to her outrage.

The next morning she showed up and didn't seem at all surprised that Sidney and I were still on the payroll. In fact, she was even cooperative, asking to see the choreography again and suggesting to me only that perhaps the background to "Show Me" could be a bit more contemporary, at which point she pulled out an audio cassette from her handbag and played us a little background rhythm she'd worked out at home with another accompanist.

"That's wonderful," I quickly told her. "Let's do it that way."

Our pianist hadn't heard Raquel's cassette and when he came in to start rehearsal, he began playing the song the way we had worked it out the previous day. Raquel didn't hear any difference, so we continued with our original accompaniment and the subject never came up again.

The simple steps Sidney suggested gave her some trouble, however. She began to analyze each one. "Now do I put my foot like this on the count of three, or turn it this way?"

"Just walk, dear." Sidney was getting impatient. "If you can just walk, we'll be fine."

It took us most of the day to get eight bars down to where it might be called acceptable when Raquel rushed over to Greg's office to again demand that Sidney and I be fired.

"Sure. That's a good idea. They're no good." Greg told her he'd take care of everything.

The next morning we were all back together again and once more she didn't seem surprised to see us.

We began to realize that she was terrified. She wouldn't admit it, of course, but she was not quite ready for prime time.

"Just be patient with her," Greg kept saying.

"We passed patient," I said. "We're into basic tolerance now."

Nevertheless we coddled and praised her and told her every little thing she did was sensational. But the number was deadly dull and we thought that if we put Jimmy Stewart in it, it might have a chance. He could be one of the boy dancers.

"He'll never do it," Raquel said.

But we knew Jimmy. He'd think it was great fun. I asked him to come to the rehearsal hall and take a look at it. Before I could explain it to him, though, Raquel decided she should be the one to tell him. He loved the idea and it took about half an hour to show him what to do.

That out of the way, the next problem would be her prerecord. Anticipating trouble, I primed everybody in the control room as well as Les Brown and his band to rave and carry on over her first take. She would think her public approved. She did a pretty good job on that first take and asked to hear it back.

Everybody came through. "Terrific!" "You should make an album!" "The best I've heard all season!"

She was so flattered that she decided to accept that first take, even though I would have liked to have tried for one more. The gang had been so convincing, though, and Raquel was so happy, I decided to let well enough alone.

Still insecure on taping day, however, she stopped "Show Me" a couple of times because she wasn't pleased with something she was doing. This time it was Greg who lost his patience.

His voice resounded on the P.A. system to the entire studio, audience included: "I'm the director here, Miss Welch. I'll decide when to stop the number."

Dean had been paying close attention to all this on the monitor in his dressing room. After it was over, Greg and Jimmy walked in to find a smiling and sympathetic Dean.

"Get rid of the broad!" was Dean's greeting.

"I don't know, Greg," Jimmy added in his familiar stammering style. "Raquel, uh, did that m-m-movie with Dean and m-m-me a couple of, uh, years ago. You, you-uh know something? She's a real, uh, *c-c-cunt!*"

Greg and Dean were on the floor.

Well, what the hell! It was our last show of the year. Everybody was eager to get out. As for "Show Me," Greg saved it by shooting

lots of close-ups of Jimmy's remarkable reactions.

There was an end-of-the-season party in
one of the other studios that Dean didn't attend but Raquel did. She was charming, for-
giving, even grateful. She seemed to realize
at last that we were all just trying to help.

The fact that Dean didn't show up for the
wrap party disappointed everyone, but
didn't surprise us. Our last show of the year
was easily ignored by our star.

We'd all worked hard for him, but there
weren't any thank-you's or see-you-next-
year's. It would have been nice to get some
recognition, but that wasn't to be. He never

Dean and Michael Landon

said, "Thank you very much" or "Hey, man, that was lousy." There were no pats on the back or kicks in the ass.

For myself, I never got any special privileges and always received the same Christmas gift as everyone else. I would send in my thank-you card and wonder if he ever received it. I also wondered who chose his gifts to us. He certainly didn't do it himself. Although they were delivered *en masse*, the presents had some thought behind them, a touch of class.

One Christmas I found a moment during orchestra rehearsal to thank him personally.

"Glad you liked it, pal," he said. "Jeanne

Raquel Welch and Jimmy Stewart

picked that out." Gosh, he actually knew that. Or was it a pat answer?

"Well, it was one of the nicest ones I've ever seen."

What I wanted to say was, "It isn't that it's one of the nicest ones or that it's rare or expensive; it's just that it's the same gift that a hundred other people on the staff got." Was I no more important to him than that? How about something special? Maybe a card! Some little recognition.

I told myself it didn't matter. What was important was that I knew he trusted me.

Years later I found out how much. "You know," Mort Viner confided to me, "every time Dean walked out on that TV stage he owed his life to you!"

Looking back, I can see how much he depended on me. As he worked on stage, he'd look around to see where I was. Nothing obvious, but if he knew I was there, just off-camera, he could relax, and know everything would be all right.

I used to wonder what he did all week while we were busy trying to make things easy for him at the studio. Was he working out in his basement gym? Sitting in his den sipping a cool mint julep? Watching the soap operas? Jumping a tennis net to congratulate a fellow swinger? Working on his chip shot at the Bel Air Country Club?

It was better I didn't know what he was doing. Certainly the show was better if Dean didn't know what *we* were doing.

I wondered if he ever mentioned the show as he left his family Sunday mornings to go to the studio. There wasn't much he could tell them. He hadn't the slightest idea of what the day had in store.

I knew that he lived in a huge red brick mansion just off Sunset Boulevard in Beverly Hills, that his beautiful wife, Jeanne, was well-liked by everybody in the business, and that she was considered a good mother.

It was obvious, too, that Dean was a good father. In those days, no matter how rich or important you were, it was difficult for a father to get custody of the children in a divorce case. When Dean divorced his first wife, Betty, the courts initially gave custody of their children to her. Later, they reversed the decision, declaring Dean the better parent and granting him custody. There was no publicity about it. Dean also had the good taste to keep from the press the reasons why. Something about Betty's drinking problem, they said.

I knew Jeanne was a classy lady who'd been an Orange Bowl beauty queen some years earlier. I'd seen pictures of her in the papers. She was chic, petite, and had the slightest air of shyness whenever she was photographed with Dean. It was as though she knew who the star was in the family and wanted Dean to take all the bows.

Jeanne never showed up at the studio. The papers said she wanted it that way. "Let Dean have the spotlight he deserves."

One thing she couldn't come to grips with was Dean's drinking image. "He comes to dinner sober," she informed the press. She also told them she was getting good and tired of all the mail she was getting from the Women's Christian Temperance Union.

"He's not a drunk!"

He certainly wasn't one at the studio. Sure, there was always that drink in his hand, but he seldom took any of it. If the level of the liquid in the glass was lower when he left than when he arrived, it was probably because he'd spilled some of it.

It was the first question my friends would ask: "Does Dean Martin really drink that much?" Dean's own answer in a press release was the best: "How could I drink as much as everyone says and still do what I do? If I were bombed, how could I come up with those ad-libs? How could I even read the cue cards? How could I know and judge an audience if I were loaded?"

"They aren't idiots at NBC," I heard him say once. "Do you think they'd pay all those millions to a drunk?"

I was still as curious as everyone else. One night after he left his dressing room to go home I took a sip of his drink, just to see for myself. It was a mild Scotch and soda.

"I drink moderately," he'd clown to the studio audiences. "Yeah, I keep a bottle of Old Moderately around all the time."

To millions of viewers, Dean was considered literally the "toast" of prime time television. Once in the studio, Greg boomed down from the control room, "Hi, Dean."

Dean stumbled back with, "Oh, am I in town?"

But Gail Martin told me that at home her father liked to go to bed early so he could get up at the crack of dawn and play golf. Once when his kids were having a late-night party, Dean picked up the phone, called the police, and reported, "The Martins are having another loud party. Can you send somebody over there to break it up?"

Soon the doorbell rang and the cops sent everybody home. The next morning around the breakfast table, the kids told Dean about the officers coming to the house.

"Somebody called the police," Deana told him.

Dean coyly admitted he was the culprit, and the whole family roared with laughter.

Dean was quick to say that he was a simple man with a modest upbringing, that he hadn't read a book since *Black Beauty*, and had never seen a Broadway show. But that was no dummy who walked into NBC Studio 4 at 1 p.m., absorbed all that was thrown at him and walked away at 8 p.m., having totally enchanted millions of viewers.

Maybe that drink in his hand every Sunday was for a little extra confidence — just for facing the unknown every week.

I like to think that his family on Mountain Drive was just as amazed as we were that he could actually pull it off. I visualized them all sitting around the TV set every Thursday night watching what a great entertainer dear old Dad was. And he hardly had to leave home to do it. He wasn't what you'd call today married to his job, that's for sure.

And so, as for Dean's private life, all I knew was what I read in the papers. When I first arrived on the scene in 1965 it was all so idyllic. As the years went by, however, things changed for Dean and consequently for *The Dean Martin Show* as well.

Our star — the unflappable Dean Martin breezes
through another show.

During a station break, the chorus girls playfully
attack Dean.

Liberace was always a favorite guest. He and Dean played well off each other.

Another frequent guest was Orson Welles, who could always be counted on to provide a colorful presence.

Jimmy Stewart finds himself the center of a production number. Jimmy graced Dean's show multiple times.

Jimmy Stewart, Dean, and Orson Wells share a centerfold during a sketch.

277

Even with the wide variety of stars on the show, we still had some surprises, like a visit from Ronald Reagan.

Music was central to Dean's image and the show. Duke Ellington, Dean, Lainie Kazan, Frank Gorshin, the Andrews Sisters, and Tim Conway gather around the piano for a song.

Dean's guests always shared in the chorus. Guy Marks, Dean, Bill Cosby, and Liberace trade verses.

Joey Bishop, Petula Clark, and Orson Welles join Dean in the closing chorus.

279

A highlight for guests, who had to make do with stand-ins for Dean at rehearsals all week, was their moment on stage with our star. (*Above*) Dean and Lena Horne.

Dean and Joey Heatherton banter while riding tricycles.

Dean shared the spotlight every week with talented ladies (*clockwise from top*): Ann-Margret, Shirley Jones, and Phyllis Diller.

Along with the music, each show featured several dance routines. Dean dips dance legend Ginger Rogers.

Sometimes the dance even entered the realm of high art: Dean with ballet stars Margot Fonteyn and Rudolph Nureyev.

Every show packed plenty of comedy sketches, always featuring the expected Dean Martin references of "booze and broads." Dean and Rip Taylor cut up.

Dean, Paul Lynde, and Gordon MacRae joke their way through a football sketch.

Sammy Davis Jr. comes to the rescue after Dean takes a pratfall.

Dean thrills the hipsters when he and Carol Channing take the stage on a motorcycle.

Dean and Ross Martin lead the group in tormenting a put-upon Paul Lynde.

Dean and Frank Sinatra were joined by their families for a Christmas show. Gail Martin and Nancy Sinatra sing along with their famous fathers.

Dean and daughter Gail share a moment on stage.

Pairing Dean with his beautiful female guests was a staple of the show. Dean and Petula Clark perform a moving duet.

Elke Sommer leads Dean through a rousing number. Because he didn't attend rehearsals, Dean often had to be forcefully guided in his dance steps by his guests.

Charles
Nelson Reilly,
Dean, Gina
Lollobrigida,
and George
Gobel belt out
a finale.

Though rarely dressed in anything less than one of his beautifully tailored tuxedoes, Dean could carry off almost any look with style.

Dean was frequently joined onstage by the Golddiggers, a group created as part of a summer replacement series to keep a hold on Dean's TV time slot. They became fixtures on the variety show as well.

Another of Dean's chorus cuties were the Dingaling Sisters, a select group of Golddiggers. Dean's Dingalings provided a sharp contrast to those of guest Jack Benny.

The inimitable Dean Martin with 1973's Golddiggers.

7 The Fifth Year — (1969–1970)

Sultry Ladies, Stormy Scenes, and a Shocker

It was back to booze and broads for Dean's fifth season, and the NBC censors hung around the studio like vultures at a fat farm to make sure we didn't go too far.

"Okay, we know it's Dean's image," they kept repeating. "We know it's a ten o'clock show . . . but there are limits." So what did we do? "Make up more raunch than we would ever want ourselves," Greg instructed the writers. "We'll give 'em plenty to cut so they think they've done their job." The ploy worked every time. We wound up with more suggestive material than we expected, yet always within our own home rules of good taste.

Over the hiatus, Mort Viner suggested I make a cassette tape of the songs Dean

Dean, Carol Channing,
and Mike Connors

would be involved in that week, so that Dean could study up a bit before showtime. Each week from that year on, Geoff Clarkson and I would record all of Dean's music. Geoff would play it on the piano, I would sing Dean's sections, and Melissa Stafford would perform all the other guests' parts. It was important that Dean understood that when he heard me singing, that would be his part. Melissa's voice meant it was somebody else's section. And of course he never knew who that would be until he got to the studio.

Although we kept doing this week after week, Mort Viner later confessed that Dean rarely listened to the tape. Maybe he'd play it on his way to the studio, but usually he heard his music for the first time when he stepped on stage for orchestra rehearsal.

As we tried to settle down again, there was a certain amount of anticipation, all of us coming back like the first day of school, loaded with new ideas we thought would be breathtaking. Still, Greg had to remind us that we were still dealing with a star with whom we could go only so far because he refused to show up until the last minute. The first day always included a lecture from Greg about how things last year could have been better, should have gone easier for Dean, or might have been cleverer and funnier. All his ranting and raving, though, was merely to get our juices stirring. When all was said and done, Greg made it easier for us by his straightforward approach to the situation.

By this time Greg clearly knew how far to go with Dean and the censors, and he was one of the few television producers who would give you a direct answer when you asked a question. If something could either go this way or that way, he'd choose one or the other without hesitation, tell you why, and stand by his decision to the end. And the few times he was wrong he admitted it.

Movie and theater people are quick to put down television because of its "imperfection." It's true — on our show there really never was time to make things perfect, try out material, hone it down until it was perfect. There are no out-of-town tryouts or sneak previews in television. But we all thought that was the great challenge, to constantly strive for perfection.

The Dean Martin Show never remotely approached that goal, but we felt we had something approaching "perfect television." We had one very unusual and ingratiating ingredient: Dean, now the epitome of virility and versatility.

With so much material to fashion for our phantom star each week, we still zipped through a season so fast that we could barely remember who or what was on last week. Sometimes things wouldn't go quite the way we'd planned, and all of us would say, "Oh, why didn't I do that?" or "Why didn't Dean react to *this?*" But it was always too late. The show was in the can and it was on to the next week.

Since we had been able to whip off two shows a week at times with no real pressure on us or Dean, Greg decided to start right out this year with two shows a week on a regular basis. He felt, though, that it might be easier for all concerned if we had two staffs — a set of writers, a choreographer, and dancers for one show, and an entirely different gang of the same for the second. He

Olivia Newton-John and Dean

and I and a few select assistants would work all shows.

Bob Sidney stayed on as choreographer for what he insisted on calling "Group A" with Ed Kerrigan as his assistant, while Jonathan Lucas was hired to stage "Group B," Tommy Tune his second in command. Both shows were set up at the same time, Greg and I jumping from one rehearsal hall to the other, trying to keep an eye on everything and at the same time attempting to soothe the tempers of any rivalry that developed between the two groups like "Our show was certainly better than *theirs*" and that sort of thing.

The confusion only lasted nine weeks, however. One day Greg arrived at the office and calmly said, "Why are we trying to kill ourselves? What's the big hurry? Have we forgotten how to relax?"

Of course it had all been *his* idea. We were glad he realized that and happy to get back to a normal one-show-a-week schedule. The two groups remained, however, working alternate weeks and rarely crossing each other's paths.

Tommy Tune did a couple more solos on *The Dean Martin Show* and tried helping the Golddiggers with their staging, but his heart wasn't in it. He was tired of television, of its imperfection and its "settling" for less than Tommy thought was necessary to make both him and the girls look their best. He hated the limitations and the lack of time to do anything right. At one point he said that director Ken Russell had seen him in England on a *Golddiggers in London* show and asked if he'd be interested in a small part in his upcoming film version of the '20s musical *The Boy Friend*. Greg made it my job to convince Tommy that he was stupidly throwing away his career and that he should stay in Burbank and be the Golddiggers' choreographer, that nothing in London or on Broadway could possibly be a better deal than that. The Russell movie, Tommy insisted, was a stepping stone to something big on Broadway.

My pleading and Greg's continuous prodding fell on deaf ears and Tommy flew off to London.

"I'm sorry he's leaving, but that kid's loaded with talent," Greg conceded. "He'll be a big Broadway name. Wait and see!"

The Dean Martin Show and *Laugh-In* used the same Studio 4 and Rehearsal Hall 1 at NBC. It was a family affair, both shows sharing staffs and getting along famously together. We all agreed that Goldie Hawn would be the perfect guest for our first show of the new season. We were sure that Goldie's refreshing qualities were just the "something different" we needed.

Dean and Goldie Hawn

Goldie seemed excited about working with Dean. She joyously read through her sketches, which also included Dom DeLuise and Dennis Weaver, both old-timers with us by now. They, in turn, couldn't wait to play television with this charming new personality.

Dom said he'd do anything with Goldie — walk-ons, bits, extra work; he just wanted to be near her. We kept giving her more and more to do, limiting Dom and Dennis to some extent. But they knew we had someone special and they went along with it.

Although we had a few musical moments built in here and there, I hesitated suggesting she do a song-and-dance number all by herself. That sort of thing didn't exist on *Laugh-In*.

She was reading my mind.

"You know what I've always wanted to do?" she asked one day. "One of those big production numbers you do on your show. Like Cyd Charisse or Ginger Rogers. Golly, just think of it! Sequins and feathers and stairways and lots and lots of boy dancers . . ."

Why not? We decided to take Goldie's wish and make a big production out of it. "Let's give her the works!" Sidney insisted.

Everything but "lots and lots of boy dancers," I cautioned. She'd have to settle for just four of them. But we'd throw in all the girl dancers, the singers, and even Dom.

The song: "The Lady's a Star!"

"It's so *me!*" she shouted. She worked long and hard on it, singing and dancing and looking every bit the "fan-to-star-in-five-minutes" we aimed for.

Dean and Dom DeLuise

Ordinarily Dean didn't know or care who his guests were each week, but he happened to hear that Goldie was booked for our first show and asked Mort Viner to get him to the studio early that day.

"Where is she? Where's Goldie?" asked Dean. We hadn't heard anything like that in five years!

We carried on our usual schedule, assuming Dean would hide out in his dressing room most of the day, but no! He wanted to be onstage every time Goldie was there. To hell with the football game; this was something special. He wanted to rehearse the sketches with her, sing all the songs during camera rehearsal, and just hang around. Unheard of!

We were worried that so *much* rehearsal on Dean's part might cut down on the spontaneity we had striven so hard to achieve. But no, their sketches were bright and cheery when we got to taping them, mostly because neither of them chose to pay too much attention to the cue cards.

"I think it's so cute the way you're always pretending to be so dumb," Dean threw in.

"I'm not pretending," Goldie quickly corrected him. "I really am dumb. What's more, dumb is beautiful."

"You're really proud of being dumb?"

Goldie returned temporarily to the script. "Dean, what's the major problem concerning our country today?" she read from the cards.

Dean decided to have none of that. "There's a problem with our country?" Goldie had trouble going on.

"Well, there's strife in our universities," Goldie giggled. "Don't forget that." More giggles.

"Why don't you cheer up?" Dean threw in.

Goldie got back to the cards for a moment. "And who goes to the universities? The *smart* people."

"Gee, I never thought of that."

"That's because you're dumb like me."

The critics knew neither one of them was too dumb:

"Hawn and Martin have similar breezy styles, an approach so casual seeming as to be Sure Death in less practiced hands. Neither can seem to do anything wrong, however, and even the occasional goofs come up comedic roses."

Dom fell into that category, too. Almost a regular now, he was one of the few performers who could remain unfazed when Dean would go out of character in the middle of a sketch.

Dennis Weaver, not known for his music or comedy talents, nevertheless made the most of those moments with Dean. He fancied himself a singer of reasonable prowess and just about floored us when he confessed that he, like Milton Berle, had always wanted to play *The Music Man*.

Once again it was Greg who furthered the idea along. Because of Milton's struggle, I desperately tried to suggest other show-stoppers. This time Greg said he knew Dennis could do it *because* Milton could.

To make it easier for him, I tried to simplify "Ya Got Trouble" right from the first day of rehearsal by letting Dennis recite some of the long patter, without rhythm accompaniment. To put it into sections so he could take a breath occasionally, I brought in the singers and dancers to do *"Ya got trouble, right here in River City, trouble with a 'T' and that rhymes with 'P' and that stands for pool"* early and often. Dennis welcomed the changes, but still couldn't tackle the long recitative.

"It was the toughest thing I ever tried to do in my entire career," he later confessed. "I never should have let Greg talk me into it. It was sheer torture."

My tinkering with the score drew the wrath of *Music Man* composer Meredith Willson, whom NBC had contacted for permission. He was outraged and refused to allow us to make "such drastic changes." Greg told NBC we'd put it back the way Willson originally wrote it. But Dennis saw nothing but disaster ahead, trying to relearn something that was already giving him migraines.

There was no way we could change it, so

Dennis Weaver

Greg took the dangerous way out. "Aw, tell 'em we're doing what they want, but we'll do it our way. Just forget they even called."

Even with all this going on, Dennis managed an entertaining performance on taping day, thanks mostly to the energy of the singers and dancers. Strangely, we never heard from composer Willson again. Apparently he didn't see the show. And NBC wasn't about to make any waves.

Other composers rallied to our cause, however. Whenever there was a need to change one of Ira Gershwin's lyrics, I would call and read them to him. He was always gracious when I explained that the only reason I needed to change them was to fit a specific situation on the show.

"I understand," he'd say. "Why don't you

read me what you've done?"

He always expressed approval, except for a couple of admittedly shameful times I neglected to make a perfect rhyme, like rhyming "mine" with "time," for example. That was something Ira couldn't aesthetically let pass. He'd think for a moment on the phone, then come up with some lines of his own, all quite superior to mine. On those occasions, the great Ira Gershwin was working for Dean Martin.

Irving Berlin was a fan, too. In fact, his publishing company claims we were the only TV show to which he would give automatic permission to alter his lyrics. On a show in which Dean and Orson Welles were the only ones available for a finale, I thought of the Berlin song "Only for Americans" from *Miss Liberty*. His publishing book said it was "restricted," meaning no one could use it. I decided to call Ben Gilbert, head of Berlin's office in Hollywood, to see if there was any chance. He said he'd call Mr. Berlin. Five minutes later there was an answer.

"Mr. Berlin says 'anything you want' and that he loved the way Dean sang 'Always' last week. It's his favorite song, you know . . . wrote it for his wife. He also says Dean sings 'White Christmas' better than anyone." I wonder how Bing Crosby would have reacted to that.

I told Ben I'd pass that all on to Dean, but naturally there was no time for such insignificant (to Dean) accolades. He probably would have said, "I didn't know he wrote 'White Christmas.' "

More important was our star's mood when taping time came around. To keep him alert and at the top of his energies after he slid down the pole, the girls were rounded up in the scantiest of outfits to gyrate erotically just off-camera during his opening song. Sometimes one of the girls might even expose a breast if that would pop Dean's eyes open, anything was fair game to get him ready for the rest of the show.

During sketches Greg stood just beside the cue cards, knowing that's where Dean would be looking. Greg would act out all sorts of emotional reactions to comedy lines as they came up. Instant direction. To the rest of us on the set, it was as much fun watching Greg clowning around as it was to watch what Dean actually did with the sketch. Greg was a performer himself — happily playing Dean for our star's benefit.

Because the show was becoming easier and easier to do, it was decided that we

Ann-Margret and Dean

could take time for a couple of specials with Gene Kelly, using our staff as well as the Golddiggers and our singers and dancers.

Dean's show seemed to be something we did in our spare time. Yet once in a while we were jerked into reality by some sparkling guest stars, like superstar Ann-Margret.

Aside from her being a sexy top name, we all knew she had been seen frequently entering Dean's dressing room without husband Roger Smith. Some questioned those visits, especially since the dressing room "regulars" came out when she went in. Her appearances in our hallways even before she guested on the show raised a lot of eyebrows.

"Remember when the *Enquirer* said she

was threatening Dean's marriage?" some-body pointed out.

Yet in my view, they were simply very good friends. They clearly adored each other. On the show, however, her "vamping" of Dean easily outdid anything the Gold-diggers had ever attempted. Sometimes it got so hot that Greg had to photograph their duet from the waist up. Those slithering hip and leg movements would never have passed the NBC censors.

Rumors aside, they did leave the studio separately.

Sexy ladies were the norm, and we booked as many as we could. German actress Romy Schneider, unlike fellow countrywoman Elke Sommer, had no sense of humor in English. She could never figure out where the laughs were supposed to be in her sketches with Dean and Charles Nelson Reilly. We painstakingly explained each joke, asking her to change her inflections so that her lines would come out funny. She was in a state of complete confusion until taping, when the audience was there to pro-vide the anticipated reaction.

It didn't help that, like with so many guests on our show, Romy didn't get to meet Dean until just before the sketches were

Dean, Romy Schneider, and Charles Nelson Reilly

ready for the cameras. Greg, right at Dean's elbow, gently led him to the set and kept the formalities brief. His introduction to Romy wouldn't have pleased Emily Post.

"Dean, this is Romy Schneider. Romy, meet Dean Martin. Okay, roll tape!" Greg never wasted time, money, or words.

He never should have insisted on a song for Romy, either. Although the voice quality was pleasantly sultry, she had absolutely no sense of rhythm. She kept Les Brown busy trying to keep up during her uneven rendition of "Ain't Misbehavin'," even with me standing off-camera begging her with huge hand gestures to "COME IN NOW!" or "NO, NOT YET!" We finally dispensed with the orchestration and Geoff Clarkson, our pianist, accompanied her as best he could on his own.

308

Carol Channing and Tommy Tune

No such problems with a pro like Carol Channing, though. Dean murmured that he was skeptical, always considering her too Broadway for his taste. But they got along famously.

My only concern was that he might pass out when he first met her. Carol, a noted devotee of strange diets, was in the midst of a new one. She ate nothing but raw fish, which she carried in a big jar, constantly sticking her hands in for a quick nibble. I was afraid she'd bring all of this onto the stage. Since for the past week we'd all been very aware of the not-so-faint aroma reminiscent of performing seals, some wise person took the precaution of placing a bottle of Listerine in her dressing room. Carol got the message.

Actually she was so much fun and so quick that we kept giving her more and more to do, in and out of sketches and songs with Dean. One was a bit on the stairs at Dean's home base. They were discussing Carol's next number:

"But Carol, I know those NBC censors — they'll never let you sing 'Love for Sale.' You know, the lyrics . . . they're too suggestive. Maybe if you substitute another word for 'love.' "

"Substitute a word for 'love'? Like what?"

" *What's* a good word," Dean replied. He started to sing.

"What is this thing called love?"

"Wait a minute," Carol giggled. She abandoned the cue cards, turned to the camera, and, flabbergasted, told the audience, "He doesn't come to rehearsals, he doesn't take the script home, he doesn't even read the cards!"

Finally, Carol's substitute for 'love' was, appropriately:

> *Fish for sale*
> *Appetizing young fish for sale*
> *Fish that's fresh and still unspoiled*
> *Fish that's only slightly soiled . . .*
> *Fish . . . for sale!*

Eva Gabor was typical of the nonmusical

Dean and Eva Gabor

guests who had never been on a variety show before but said, "What the hell, let's have a crack at it!" Eva couldn't carry a tune, but unlike Romy, she knew it.

"What can I do, dahling? I never sang in my life."

It didn't matter. She had a good sense of rhythm, knew how to get the best results from a comedy line and, of course, had that wonderful Gabor accent.

Her older sister Zsa Zsa was not so tolerant of us, though. After suffering through a whole week of my doing Dean's part, Zsa Zsa finally exploded on stage just as Dean was about to watch her sketch in his dressing room. She had thought that now at last she'd be working with her costar. But no, this time instead of me she got Greg, who had thought it would please her to have the producer standing in.

Greg was totally unacceptable to the Hungarian diamond collector.

"When am I ever going to hold the hand of the man who is supposed to be in this sketch?" she screamed. "I can't work this way!!!"

Neither could Greg. He responded to her temper tantrum in his usual blunt way. "Okay, lady, get your ass off this stage! We won't miss you!" She stormed off, muttering some obviously choice Hungarian words under her breath.

Dean, having watched this tender scene on the monitor in his dressing room, couldn't resist coming out to tell Greg, "It took you two minutes longer than I predicted it would!"

Greg took a chance on filling Zsa Zsa's place with a pretty girl singer who'd been booked to do a short spot on the same show, a slim, shy blonde from Australia with a less pronounced accent than Zsa Zsa's, Olivia Newton-John. It was her introduction to prime-time American audiences.

Occasionally Greg would take an agency's advice about an unknown. Olivia was wholesome, pretty, and engagingly untainted by

William Conrad, Olivia Newton-John, and Dean

show business. She'd be good with Dean, we agreed, if only to just stand there and let Dean work his usual gentlemanly charm on her.

Her voice seemed almost too tiny, too tinny — but that face! It was easy for Greg to go in for some extremely tight close-ups and let the microphone be close enough to capture her breathy tones.

When told she'd be replacing Miss Gabor, she was both excited and frightened. It was too late to put her in the sketches, but she easily replaced Zsa Zsa in the finale.

Greg took care of the earlier unpleasantness with his usual flair. He sent Zsa Zsa a big bouquet of roses. She called to apologize and the next time she did our show she kept her famous temper under control.

Two ballet stars, Rudolph Nureyev and

Margot Fonteyn, seemed about as far removed from the Dean Martin image as we could imagine. However, their agent said they'd love to do the show. How odd, we thought, but on the other hand, why not? Maybe we'd pick up a whole new audience.

We wisely scheduled them early on taping day, pushing the rest of the show back a few hours to accommodate them. They'd definitely need extra camera rehearsal.

We invited an audience of dancers and other interested parties instead of the usual bus tours from Pasadena.

Rudy showed up late with a terrible hangover and we couldn't get him out of his dressing room. His embarrassed manager tried desperately to sober him up with coffee and pills, but Rudy decided to remain temperamental. None of Greg's shouting could budge him. The crew, the audience, and the cast waited.

We shot some scenes around him until he finally ventured out with Margot to test the stage.

"Impossible!" he shouted. "I'll never be able to keep from falling." Because of last night's party, we thought, not the stage. The crew, nevertheless, brought in a whole new platform for him to judge. Then Rudy decided to have lunch!

Normally a scene like this would have simply meant the performer would be asked to leave. But Nureyev and Fonteyn were special, and Greg felt their appearance on the show would add a certain class. He decided they were important enough to be allowed to carry on in this unprofessional way. He called Dean at home and told him not to rush in.

The audience once again was told that "technical difficulties" made it necessary to postpone the ballet taping. Some hours later Rudy decided he was ready. He and Margot got through *Sleeping Beauty*, not to their approval, but certainly to Greg's. Rudy was not in good enough shape to put up much of an argument.

Margot Fonteyn and Rudolph Nureyev

Dean showed up as the ballet was ending and walked on-stage with a bouquet of flowers for Margot, who left to let Dean and Rudy fend for themselves. A rather awkward situation, but Greg felt that Rudy was the star of this dancing duet and he ingloriously waved her off.

"I don't know much about this bal-letty stuff," Dean confessed while twirling around Rudy.

"That's almost a pirouette you're doing." Rudy was not amused but submitted to the writers' idea of funny patter.

"Right. The peer-o-etty. That's good for screwing lightbulbs in the ceiling."

Rudy was happy it was all over. And so were we. He made an unballetic exit towards his dressing room, screaming at his agent for letting him get involved with all this.

Even though Dean had been part of the so-called Holmby Hills Rat Pack (which included Frank Sinatra, Peter Lawford, Humphrey Bogart, and others) we wondered why still another member of that infamous group had never been a guest on our show. Supposedly Sammy Davis Jr. had been one of Dean's closest friends during that time. The Pack had fallen apart, of course, but shouldn't there have been some recognition of Sammy now that Dean had his own show?

An oversight, maybe. Dean never made any suggestions about guests one way or another, leaving that all up to Greg. To our surprise, Sammy showed up on one of the shows that fifth season, finger-snapping his way through rehearsals all week and sporting a Nehru outfit loaded with necklaces and rings. Maybe, we thought, Dean had outgrown this early '60s stuff and Greg felt Sammy would be an embarrassment for Dean.

Dean's first comment on taping day certainly bore out this theory.

"When does he do his war dance?" he mumbled to Les Brown's band.

I'd put together a long medley for them, full of songs Sammy was known for. But it just lay there. Greg's antics off-camera didn't help. Even Dean looked as though he thought *something* had to be done. So what did he do? In lackluster situations, he always did something unexpected. This time he fell on the floor.

Dean had an inner sense of trying something outrageous to liven things up. He might merely collapse, fall off a stool, or purposely tangle up his feet and slip. This

Dean and
Rudolph Nureyev

time it awakened the audience and broke up Sammy. But then, in those days, *anything* broke up Sammy.

In spite of all the star guests, the look of the show was still booze and broads . . . until the night before our last taping for the season.

It had been on the radio news that morning. Yet nobody on the show really believed it: "Dean Martin and his wife, Jeanne, are splitting after twenty years of marriage!"

It was a bombshell. Yet we'd been warned. She had seldom come to the studio, that was true, but that was Jeanne's way. An ex-performer herself, she didn't want to interfere in Dean's professional life.

When we finished taping that day, we

found the news was definitely true. They were getting a divorce. If Dean had been going through domestic turmoil, though, it had gone unnoticed by us.

Now, since this turn of events had been verified, there was work to do. Dean had made Jeanne a part of the show by constantly referring to her, and there were ten more shows taped that hadn't been aired yet, all with "Jeanne jokes."

Greg talked to Dean about editing out those references, questioning the taste of leaving them in. Dean agreed it had to be done.

We waited until everyone except the necessary crew had left the studio. The doors were locked. Dean sat on his famous stool while the girls on the production staff scanned through the scripts of the last ten shows. For a man who usually hated to wait, he sat patiently, reading each new line as the writers submitted them. He showed no emotion and tried very hard to imagine a laughing audience reacting to the new material.

After four hours, Jeanne was erased from the show.

The public was waiting. The following week on the air, we got our highest rating ever. Audiences expected to hear Dean say something about the separation. When he didn't, the next week's show got our lowest ratings.

A psychologist might point out that the audience believed that even though Dean was forever touching and talking about girls on TV, he was really only kidding. Dean's viewers knew he was a good husband and father and the booze and broads jokes were just part of the act. When they heard about the divorce, though, they might have said to themselves, "Maybe he wasn't kidding after all."

Yes, we had a lot of girls on the show. There was constant temptation. Some of that temptation came from a scary source, the mob. Through the seasons, the show did a lot of favors . . . for "the boys" from Cleveland who knew Dean, "the boys" in Chicago who knew Greg, and "the boys" from New York who knew them both. The favors were always the same. For instance, there was a blonde singer who had just finished three days at the Latin Quarter. Now she was opening at a joint in Cleveland. Could we put her on the show? Of course we could. Sometimes their songs would make it to the final tape; sometimes we would "run out of time." I was stuck with putting the musical minutes together with the girl and Dean. Usually, it aired.

Greg also did a favor for one of NBC's biggest stars, big on radio, bigger on television, a beloved comedian. He called and asked if we could use a "friend" of his on the show. Greg agreed. In fact, she did bit parts for a couple of weeks. She was only nineteen, a beauty, and when Dean's divorce hit, the New York newspapers picked up the story. They insisted the reason for the divorce was this 19-year-old girl. Dean knew who she was and who her "friend" was. He took the rap, never mentioning that she was the personal property of that NBC star.

One of our show's dancers, who had had dinner with Dean a few times and thought the divorce story was about her, immediately started asking for her own divorce so she could marry Dean. Greg set her straight.

So which one was it? That other dancer who always managed to get closer to him than anyone else? One of the Golddiggers? Ann-Margret? Petula Clark?

Actually, it was all of the above. And probably more. "He was a naughty boy," Jeanne sighed.

The public was not kind to Dean. They suddenly saw him as a "dirty old man."

The Dean Martin Show would never be the same.

8 The Sixth Year – (1970–1971)

Picking Up the Pieces

We made *The Golddiggers in London* shows in the summer of 1970. Four weeks after that experience we were all back in the sterile halls of NBC-Burbank, remembering with a sigh the English tea ladies, rehearsal halls with a view, Elke's Lady Godiva imitation, and waiting for the start of Dean's crucial new season, year one A. J. (After Jeanne).

Although most of the reaction to Dean's marital problems had died down, Greg decided to take no chances. Dean was going to have to look like a saint compared to those first five seasons.

His famous pole was replaced by a nice, slow elevator. He was accompanied on the ride down by only one girl, daintily dressed, who would disappear as soon as she had positioned him at his stool for his first song. His monologues avoided any booze and

Dean and Pat Crowley

broads jokes, dwelling mostly on current events.

The biggest change was the closing. Instead of those wild and woolly burlesque blackouts with their raunchy jokes and scantily clad girls, Dean sat in the middle of a huge pink set in an Andy Williams-style sweater, the Golddiggers perched around him at safe distances.

He sang three or four songs, mostly pretty ballads, always beginning with "Welcome to My World." We called it our "closing concert," hoping to establish something safe and respectable. The girls were dressed very primly and there was no touching, leering, or naughty lyrics. See Dean sing. See how warm and innocent the girls are. See how clean and perfect we all are.

Without sex and shady jokes, Greg figured we'd better throw in a kitchen-sink-full of big names to pique the viewer's curiosity. Petula Clark was on hand, and in her multichorus version of "Call Me" the writers came up with one-liners for the likes of Orson Welles, Joey Bishop, Eva Gabor, Tom and Dick Smothers, Dom DeLuise, Dionne Warwick, Lee Marvin, Dan Blocker, Joey Heatherton, Alan Sues, Sugar Ray Robinson, Glen Campbell, Dick Cavett, Pat Crowley, Merv Griffin, Johnny Carson, Flip Wilson, Lloyd Nolan, Vince Edwards, Paul Lynde, Raymond Burr, Mike Conners, Carol Burnett (she taped it at her studio as a favor), Ernie Borgnine, Arte Johnson, Danny Thomas, Charles Nelson Reilly, Rowan and Martin, Lucille Ball, and Bob Hope. Each one popped in during one of the "Call Me" breaks. None of them were advertised in advance. "Just a nice little surprise for anyone who watched our first show," Greg announced.

Red Skelton had just shifted from CBS to NBC so he agreed to "come out of Dean's closet." That was a gimmick we'd used from the second season to surprise Dean in the middle of the show. It was always genuine. He never knew who would be in there. The "mystery closet guest" was kept secret to everybody on the set but Janet Tighe and

Greg. Even I wasn't told who it would be. It meant we all could be surprised along with Dean. The "closet" celebrities were sneaked in and kept hidden away until the last possible moment for the full effect.

Red appeared on that show as his slightly inebriated character asking, "Is this *The Flip Wilson Show*?" "No," Dean answered.

"That's funny, I thought I heard people laughing in here." Red presented Dean something in a brown paper bag. "This is a Willy Lumplump Special — half a pint of gin, half a pint of vodka, half a pint of bourbon, with a black widow spider swimming around on top."

"Won't that make me drunk?" Dean inquired.

"It'll help!"

The critics were impressed with everything about the show:

Daily Variety:
What Greg Garrison put together to blast Dean Martin into his sixth season could serve as a text for producers of shows for the family. Also a model for a musical or a comedy. What's missing wasn't missed. It was that altogether, the kind of show to relax to and enjoy. Only apparent change from the

past is that Garrison has called for more of Dino's patent singing. More than 30 stars (the best in the land) were spotted for cameos as a gag for Petula Clark's "Call Me."

Studio laughs submerged much of the punch lines. The laugh meter never got such a workout.

Weekly Variety:
To be sure, Garrison had some yeoman help to field such a fiesta, some ten writers. Les Brown rates a bow for the music backgrounds. But the combination of Martin and Garrison is what puts the big numbers on the board. The closer, in place of "Keep those cards and letters coming in" sign-off had Dean pleading, "Let's work for peace, not fight for it." There was never a dull lag in the show and how many others can make that statement?

Greg still worried about the ratings and Dean's image and decided to load up the show with names every week, like in those early Hal Kemp days. Our second show read like a talent agency's big gun list: Peter Falk, Shirley Jones, Paul Lynde, Kay Medford, Kenny Rogers and the First Edition, Jackie Vernon, Concetta Colaleo, Lancelot Link and the Evolution Revolution, and Joe Namath.

Dean and Ruth Buzzi

Peter Falk turned out to be a nervous actor who, although he said up front that he understood the way things were done on the show, couldn't get comfortable with me as Dean's substitute during rehearsals.

When Dean did arrive, Peter couldn't resist telling him what a trial it had been. "I've been busting my ass all week while you're out there on the golf course, and it ain't been easy. Couldn't you have found a stand-in who could *act?* Just a little?" Part of his tongue was in his cheek when he said it, but I knew he'd been not too bravely putting up with me all week.

Peter was also unprepared for Dean's ad-libs, and it threw him here and there. "Ain't we gonna start over?" he kept begging Greg. No, Peter. Carry on, Peter. This is *The Dean Martin Show!*

★ ★ ★

We had managed to stay away from male singers that year, preferring to let Dean handle the ballads exclusively, and allowed the actor guests to do what they could in the musical comedy sketches we devised. But along came a singing star who Greg felt would work well with Dean, Englebert Humperdinck.

He sauntered into our first rehearsal and immediately wanted his "Big Ballad" to be part of the show. He and his manager fought us bitterly when Greg insisted on an up-tempo fun number instead. After getting nowhere, Greg dropped it in my lap. I had watched the bickering and decided to take the old flattery route.

Dom DeLuise and Ruth Buzzi

"Listen, Englebert, your voice is so much bigger than Dean's, so much richer. You're younger, you're taller, your records are selling better than his. Let's face it, you're a threat! I'm sure you can appreciate our problem. It's Dean's show. We can't take a chance that you'd show him up."

I knew he could never do the latter, but he fell for the blarney and agreed to do one of his lighter tunes. He spent the rest of his time in rehearsal going around explaining to the cast and crew why he wasn't doing a big ballad. "I don't want to make Dean look bad."

Dean, of course, couldn't have cared less whether Englebert sang a ballad or a hymn. Or whether he even sang at all. He never thought of another performer as a threat — just someone to play with.

Some weeks later Kay Medford's character as Ken Lane's mother was expanded into a sort of sitcom in the middle of our show. Marion Mercer portrayed her daughter, Tom Bosley was Marion's boyfriend, and Abe Vigoda played Kay's constant admirer. Dean and each week's special guest made cameo appearances. It usually ran about twelve to fifteen minutes, with the hope that a spin-off might be possible if it got great reviews. Even though Kay seemed just right as the nagging mom, Greg's first choice was Bea Arthur.

Bea had been sitting in her New York apartment, vowing never to succumb to television. She had been a big hit on Broadway in *Mame* and wanted to stay there. Greg's pleas to come out and give TV a whirl were all met with polite but firm negatives.

"I don't want to do a weekly show," she said. "Or any show for that matter."

Greg said he'd settle for one or two, just as a trial run. Her answer: "I'm perfectly happy here in Manhattan, thank you. California, especially TV in California, doesn't thrill me in the least."

Norman Lear and the TV series *Maude* eventually changed her mind, but how great that golden girl would have been with Dean!

"FRANK SINATRA IS DEAN'S BEST FRIEND," screamed a headline in an *Esquire* magazine. It went on to add in equally large letters, "HE SEES HIM ONCE A YEAR." Well, whenever they did get together on our show the ratings zoomed up. So at least Frank helped his friend in that department. He had socked it to the Nielsen ratings on Dean's very first show in 1965 and again with the "family show" in 1967. It was time for another shot in the arm. Frank to the rescue!

This time it was our New Year's Eve show, and the only other guest was Ruth Buzzi. Poor Ruth. It was tough enough to rehearse all week on other Dean Martin shows without our star. On this one, she had to contend with Jack Halloran and me as replacements for Frank *and* Dean. Our two Italians waited until taping to drop into the sketch, with Ruth as her inebriated character Doris Sidebottom caught in the same hotel room as the celebrated stars. "Dean Martin and Frank Sinatra!" she screamed. "My reputation will be sullied." Greg trusted Ruth to make the sketch work. And it did, hilariously. Ruthie is the best.

Frank tried to bring Dean back to his old self, but wasn't entirely successful. We all thought it must have been the separation from Jeanne. Dean just didn't seem as happy, as loose. He even looked older. And just when we thought he was indestructible!

Here was a man who had everything. He was rich, his work was easy for him, he didn't get involved with causes like Sinatra did. Formerly the on-time champion, he began to arrive at the studio a little late each week.

Perhaps Dean was more complicated than we had thought. We knew about his aversion to illness. For all his masculinity, he trembled at the announcement of a friend's death, or turned away from discussions of sickness.

Could his own health be a problem now? Dean was the type who wouldn't readily seek medical treatment. If he ignored pain, maybe it would go away.

Greg, trying to shelter everything about Dean, made the statement that "Dean's been suffering badly from stomach ulcers." Okay, we accepted that, for the time being. Looking back over that period, we all tried to find the real answers to why some of the fun was gone. More booze? More pills?

It was just the beginning of Dean's downward slide.

The Golddiggers, all sweetness and light, had changed their image along with Dean. Pretty dresses, less makeup, and no touching. Even *TV Guide*, in its "As We See It" column, gave us reassurance:

Our candidate for male chauvinist of the year is Greg Garrison. If any ladies in the audience want to throw verbal rocks at him, that's their privilege. Our own inclination is to set up a statue of him on Avenue of the Americas in New York — in full view of all those network headquarters.

Greg Garrison is the man who produced The Dean Martin Show and he is the man responsible for the Golddiggers. For the past three summers the Golddiggers have been Martin's summer replacement. Right now they're featured with him — largely as scenery. The loveliest scenery on television.

In the language of women's lib, the Golddiggers are still sex objects. Fully covered in high-necked, loose gowns, ungarmented to a fare-thee-well, the Golddiggers are sedately seated around Martin in a closing vocal background for him. Occasionally some of the girls do a few solo lyrics. But mostly they sit there and the camera slowly and lovingly pans from one beautiful face to another.

Women's lib or no women's lib, there is such a thing as physical attraction, and lovely, wholesome-looking women have it. It may be blasphemy to say so nowadays, but gazing at beautiful girls is a delightful pastime for men, and stage and screen movie producers have known it for a long, long time. Greg Garrison seems to be the only television producer who knows it; more's the pity.

★ ★ ★

The season was half over when Greg felt it was time to sneak a little life and sex into the show. He came up with the idea of picking out four Golddiggers for special duty. It would be a flashy quartet, returning to practically undressed luscious ladies. They'd be introduced on the Sinatra-Martin New Year's Eve show.

"Let's get back to the old days and give it a try," he said. "If it doesn't work, we've still got the Golddiggers, all prim and proper."

Dean and Frank with the Dingalings (Susie Lund, Michelle Della Fave, Tara Leigh, and Wanda Bailey)

The four girls picked to perform this little experiment were Wanda Bailey, Susie Lund, Michelle Della Fave, and Tara Leigh, that gorgeous brunette with the big operatic voice.

"Born twenty years too late," we all said about Tara. If she'd been around during the MacDonald-Eddy operetta days she would have been a big star. The Golddiggers' mail leaned heavily in her favor — everybody wanted to hear more of her beautiful soprano.

When she first auditioned for us we couldn't believe our ears. We loved it, but it was so far removed from the tinny little voices of the girls we had, and we wondered why she even bothered to show up to audition. And that name! She must have gotten that from Margaret Mitchell.

Janet Tighe and I figured there was no way she could fit in our group, but Greg saw it otherwise, especially since Tara was a stunner. After a long one-on-one chat with her, he announced that she would definitely be a Golddigger.

My God, I thought, what will I do with her? She'll stick out like a sore thumb.

"She's different," Greg confided to me later. "You can work it out."

He was right in hiring her. The mail poured in. He was also right about how to use Tara. Whenever she did a few bars with Dean, he'd give her a look of total wonderment. Those short moments did wonders for her. The public wanted more.

Dean, Elaine Stritch, and
Ernest Borgnine

I took advantage of that voice as much as I could, and now that she was a part of a smaller group she'd get even more exposure.

Here they were, the four select Golddiggers working their first public appearance on nationwide television in a spot with Dean Martin and Frank Sinatra. Greg called them The Dingaling Sisters, partly because he thought the name was funny, but also because he knew they really *were* a little dingy to be so blasé about their first shot.

On the outside, they acted like well-seasoned pros.

"I hope Dean gets it right," they moaned before the show began. The rest of the Golddiggers looked on quietly, wondering why *those* particular girls were singled out.

Having Frank on the show reminded us of

336

how well the Martin and Sinatra kids had come across on the 1967 family Christmas show. We decided to invite the performing offspring back, and add some kids of other celebrities.

Gail, Deana, and Dino represented the Martins and Frank Jr. was the only available Sinatra. Lucie Arnaz and Desi Jr. had been featured on *The Lucy Show*, so we knew there was plenty of talent there. Meredith MacRae, eldest daughter of Gordon and Sheila, not only sang well but worked in several sketches with Bob Newhart, the only other guest. Maureen Reagan, daughter of Jane Wyman and Ronald Reagan, completed our second-generation revue. They all worked in ensemble stuff, it being my job to pass things out in a reasonable and entertaining manner. No solos, just a lot of small groups, making it look like one big party.

Lucie showed the most talent, and although we felt she deserved to belt one out by herself, we didn't want to start a family feud. Maureen had a down-to-earth gutsiness, undoubtedly something in her genes. Never one to mince words, she often came to me to point out the inadequacies of the others. She also surprised me by being versed on just about any subject imaginable. She was able to jump from show business to politics to sports with the greatest of ease.

The only thing Maureen felt insecure about was her singing, so I tried to make that as easy as possible for her. "I talk better than I sing," she admitted. "Got any soap boxes I can stand on?"

Frank Jr. could easily have scored with a solo, but once again we didn't want to ruffle any feathers. He regretted that the other kids fooled around at times, and often came to me to apologize on their behalf.

He became my unappointed assistant, taking over nicely to pound some notes into the unmusical youngsters. I'm not sure they appreciated that, but it saved me considerable rehearsal time. Frank was all business, serious enough to make the others look like spoiled brats. He never hesitated to thank me for any advice I threw his way. He was easily the most professional person in the group.

Paul Lynde became a regular on Dean's show that year, working in more and more sketches and complaining more and more about every one of them. He would moan every day about the impossible situation of not working with the star until the cameras rolled. Yet he kept coming back for more. The money, I suppose, made it worth the struggle.

Andy Griffith, Dean, and Paul Lynde

Actually, as time went on he became more adjusted to the necessary trauma of never being sure how Dean would get through a sketch. I tried to play Dean in rehearsals as deadpan as possible, so Paul and the others wouldn't get used to expecting Dean to react a certain way.

Paul said he was getting to know Dean better and could pretty much predict how he'd play a scene, even assuring first-time guests not to worry, that everything would somehow turn out just fine.

"Don't be surprised," he told them, "if Dean does it perfectly and *you* goof. It's happened to me many times."

One day Paul was late for rehearsal because of a plane delay from Chicago. To save time until he arrived, Ruth Buzzi and I read

Dean and Jonathan
Winters

through all three parts in a sketch and had it
blocked and ready when Paul came through
the door. He was in a terrible mood. Not
only had his flight been late, but the airline
lost his luggage, in which he had mistakenly
put the keys to his house.

It didn't help that this particular script
was a real bomb. Ruth and I had already remarked that even Paul couldn't save it. After
listening to him snarl, we knew it wasn't the
best time to get started on this dreadful
sketch. But we had to. Paul's delay had put
us behind schedule.

He started to read through his lines and
couldn't believe how bad it was. Then to everyone's surprise, choreographer Jonathan
Lucas suddenly jumped up and yelled,
"Okay, let's put it on its feet."

"On its feet?" Paul wasted no time. "Are
you crazy? This thing couldn't make it to
Encino High."

340

Paul was mad, even furious.

Jonathan decided to more or less ignore him and plow right along. With a disgusted sigh, Paul agreed to read his lines. As expected, there were no laughs. Deadly silence. About halfway through, Ruth departed from the written word. She used a gimmick that had titillated her *Laugh-In* cast many times — blowing into her elbow, she was able to make what sounded like a fart.

She proceeded to make the breaking-wind sound with her elbow every time she felt the writers had meant there to be a laugh. Paul got so caught up with it that he actually burst out laughing. It became contagious. Ruth's fake farts, timed perfectly, somehow made the sketch funny.

When newcomers came into the rehearsal hall, we naturally refrained from letting Ruth make her revealing noises, and none of them could understand why Paul and the rest of us would break up every time we came to one of the "fart" spots. Obviously they figured it was some sort of inside joke or else they'd missed an important cue that set it up.

When it came time to do it on the set for Dean to watch, we left Ruth's noises out. Dean looked at it in his dressing room, saw us all breaking up and he, too, thought he'd missed something.

He appeared on the set and asked me, "What the hell is so funny about this sketch?"

Dean never got an answer. Without Ruth's farts, it was a hopelessly unfunny bit, beyond saving, and Greg cut it out.

We'd heard that Debbie Reynolds was anxious to do our show, yet she never appeared on any of the guest lists. Big movie star. Talented lady. Could be fun with Dean. Rumors suggested reasons why she had never been booked: Dean thought she was too butch, Greg didn't get along with her, or she was always working and never had time for us.

Whatever the reasons, one day we were told that she indeed would be on the show and we looked forward to a fun week.

For her solo, Debbie asked to do a section of her nightclub act, which Greg agreed to, even to the extent of hiring the two boy dancers from her act as well. What she didn't tell Greg was that the number she wanted to do was about ten minutes long, a good three times longer than what Greg would tolerate from any musical guest.

"I'm afraid we're going to have to cut it down," I politely told her when she walked in for the first rehearsal.

Debbie was not happy. "You'll ruin it. Don't touch my medley, you got it?"

I got it all right. It meant I had to call in the troops. As she went through the Broadway medley with her boys, it was easy to see that the last three minutes were the best part of the routine. "It's easy to cut," I pleaded.

There were some nasty words about how I was trying to destroy her number when Greg answered my distress call. This needed a lot of his well-known diplomacy. He burst into the hall in his usual flamboyant way, went straight to Debbie, and led her over to an unoccupied corner. There the two of them talked in muted tones for fifteen minutes or so, Debbie occasionally looking up to glare wickedly at me.

Suddenly Greg gave her a friendly little peck on the cheek and sauntered towards the door, pausing briefly as he passed me to say, "Make the cut!"

I never asked him what he said to her. It didn't matter. She walked over and became incredibly sweet. "Now what did you say you thought would make a good cut?"

Once again I insisted the final section was socko. She said to just do it and get on with it.

In spite of what she thought about me, she

had some very funny moments on the show, the best as a floozy in an old burlesque courtroom sketch. Paul Lynde was the frustrated attorney out to bust Debbie, and Dean was the judge whose eyes popped out every time Debbie moved. (Lots of rim shots from drummer Jack Sperling.) Dean remarked afterwards that she made the sketch funnier than it was and that his regard for her as a performer had risen considerably.

She gave it her all, but she never came back to the show.

Ernest Borgnine made several appearances on our show that year, mostly because he always made Dean smile; they had a great rapport. "I love the guy," Dean announced. "I wouldn't mind having him on every week." We all agreed. Ernie was a lovable guy, and he kept us all in good humor. He really enjoyed doing our sketches and even took on a song or two when asked. I ventured into dangerous territory again and whipped up a special-material number called "A Song for Three." It pitted Ernie up against Dean and the rousing Broadway legend Elaine Stritch. Ernie, with his charm and good nature, held his own, and the threesome was a knockout.

For the finale, I had Ernie and Elaine do a swinging send-up of Steve and Eydie that turned out to be surprisingly accurate.

"I never did enough comedy in my career," he explained. "On your show, acting is fun."

Greg decided to find out if Marty Feldman could work with Dean, which meant Marty would have to be considerably more disciplined. Greg booked him on quite a few shows, starting out with just one short bit and then pushing him into more sketches. But Dean didn't warm up to Marty. He said it was difficult to look at Marty's wild-eyed face. And Marty's off-the-wall British humor didn't exactly strike Dean as funny. So Mr. Feldman did spots without our star.

Dean and Marty Feldman

In London, when he worked with the Golddiggers, Marty refused to be a part of any musical number. "I can't stay in tune," was his reply to our requests. But suddenly after we returned to Burbank, he pleaded with me to put something together for him.

It turned out to be an outrageous ballet with Marty and two very tall, very fat ballerinas. All of this was keyed to Marty's vocal of "All I Need Is the Girl" from the legendary musical *Gypsy*. He did an amazingly nice job singing it and, of course, he couldn't help but make his usual mayhem with the dancing. At the end, the two rotund ladies graciously picked him up and flung him off the stage.

Still, Greg felt that Marty's sketches taped with Dean were not really very funny, and most of them landed on the cutting room floor.

The Dean Martin Show had another end-of-the-season party, this time a little drinks-and-potato-chips affair in one of the rehearsal halls, a far cry from the grand soirees we had those first years when the set designer outdid himself dressing up a huge NBC stage for the occasion. Cocktails and dinner were perfectly catered in those early days.

Dean sent his regrets again this time. But

then nobody really expected him to come. We'd learned. Even Greg made only the briefest of appearances before scurrying out.

There was something missing in the 1970 season. Nobody could spell it out, but we all sensed it. Dean had had the fun kicked out of him. A combination, the crew thought, of his reported ulcers and marital problems. Or maybe it was the laundering of the show. It just wasn't the same without the booze and the broads. Or the girls constantly pawing him. Or his weekly promises to Jeanne: "Don't worry, honey, I'll be home right after the show." No, it wasn't the same.

There were a lot of little things, too. Dean's hair, for instance. Oddly, the color kept changing each week. So did the styling. He'd be tousled and blond one week, combed and dark the next. All of which was tough on the editor when we had to use sections from different tapings.

He'd never been a partyer, and that year we heard he was bordering on being a recluse. He seemed to be suffering and we all felt sorry for him. We continued to respect the distance he wanted to keep from us, so there was no real help we could give him. The more the energy level went down, the more concerned we became.

Near the end of the season, someone appeared on the set with increasing regularity. A new face. She was by his side whenever he came out on the stage. We began to put the pieces together.

Kathy Hawn was an attractive young blonde receptionist in Gene Shacone's chic hairdressing salon on Rodeo Drive. She'd made no secret of the fact that she was going to "get" Dean Martin and she didn't care how. Shacone's lady customers saw how chaotic the reception area had become. Appointments were haphazardly assigned or forgotten. Everything revolved around Kathy's daily reports on how her conquest of Dean was coming along. It became the big gossip in Beverly Hills.

Over at Burbank we generally had no time for this sort of nonsense, but now it related to us. Unlike Jeanne, Kathy was showing up at all the tapings. She was always on the set when Dean was. We began to hear her comments about the show, none of which were very positive. There must have been even more said when she got him back to the dressing room, we thought.

We were concerned about those changes, subtle at that point, and we wondered if they would affect the show.

The next season we found out they would, in a big way.

9 The Seventh Year – (1971–1972)

There'll Be Some Changes Made

As we started our seventh season in 1971, Dean still owned Thursday nights at 10. New writers were hired, Bob Fletcher came on as set and costume designer, and the Dingalings took over for the Golddiggers. Jaime Rogers, who had staged the Golddiggers' first syndicated show, was signed to do the same for the Dings.

Liberace returned again, this time amid a rousing blast of headlines in Las Vegas for appearing in short hotpants. Sequined, of course. He was such a sensation that Greg saw the opportunity of taking advantage of the publicity by booking Liberace for our first show of the season. With one very strong condition: he would have to wear his jazzy new shorts.

Greg, who thought no special material was understood west of Broadway, asked me

Art Carney
and Liberace

to write something especially for the occasion. "Nobody'll be listening to the lyrics,
anyway," he had to add.

Bob Fletcher created breakaway tuxes for
Liberace, Art Carney, and Dean in which
hidden wires were pulled to drop their tux
pants and reveal a mass of sequins and hairy
legs. One bit of history was made that night.
Viewers got a rare chance to see Liberace's
dimpled knees, Art Carney's thick thighs, and
Dean's muscular calves, all on the same stage.

Much to his anger, Liberace had very
little else to do on the show. Greg had hired
him for his hotpants and that was it. No can-

Me and Ginger Rogers ready to descend down Dean's two new poles

delabra, no piano, no solo at all. Lib swore he'd never be on our show again, and he wasn't.

Art wasn't happy, either. He made no bones about his distaste for the sketches, and his harping succeeded in Greg demanding two new ones to replace the sketches Art was unhappy about.

Both Liberace and Art were pleased with the finale, though. Such a simple idea. Lip-syncing to well-known recordings. It was set up as a disk jockey show, with "Dino Vino" at the controls. Carney was hilarious in his handling of Jan Peerce's record of "Bluebird of Happiness." He and Liberace wore short wigs as the "Crew Cuts" for "Sh-Boom." In black bouffant wigs, they were joined by Dean as the Supremes in their recording of "Stop! In the Name of Love."

Dean and Liberace

We made the finale a weekly ending for the show that season, and the job of editing the best small sections of various records was mine. I bought a $1.98 audio tape-editing apparatus, and after transferring a record to tape I would proceed to edit it down to a 30-second piece, hopefully with a good start and finish. My amateur editing was only meant for rehearsal, but those little jobs I did in my office wound up in NBC's sound department and were used on the show exactly as I'd edited them. So much for big-time network television.

For years Dean had whimsically made a niche in television history by sliding down his fireman's pole, but Greg now decided to do a variation of that. While the previous year the pole had been replaced by an elegant elevator, now the pole was back. Not just one pole, but two.

"That should make it twice as much fun," was Greg's explanation.

Dean would make his entrance on the left pole, his musical guest sliding down the right one, the two of them joining center stage for the opening song.

Everybody but Kate Smith gave it a try. And Greg, who we knew could be very persuasive, almost had Kate talked into it. It was one of the few times he didn't get his way. Even though years ago she had closed her vaudeville act with a series of cartwheels (recorded for history in one of her early Paramount movies), Kate felt that at her age it was simply unladylike. Greg loved the idea of people talking about "Kate Smith sliding down Dean's pole." But Kate held firm.

"Come on, Greggie, I'd probably break it!" Too bad. It would have made a hell of an entrance!

A star who wasn't the least bit hesitant about coming down the pole was Ginger Rogers. It was her first visit with us and we were sure she had successfully found the Fountain of Youth.

As her partner all week during rehearsals, my mind kept going back to those Astaire-

Rogers musicals. Here I was, a clod if there ever was one, dancing with the great Ginger. I apologized each time I stepped on her toes, but she graciously made it easy for me. I told her Dean would be much closer to Fred than I was.

I also informed her that because Dean would be going through this routine only once, she would have to be forceful with him, practically leading him around the floor. She began practicing with me, and I found her remarkably easy to follow. I wondered if Fred ever thought that.

She was a very strong lady. She led me here, she led me there, as though her reputation depended on it. But the more I relaxed, the easier and more enjoyable it became. I knew I'd have to at least tell Dean to let her lead him around. Later, I asked him if he did.

"I felt like I was at a truck stop," he said, "but let's face it, I've always been a good follower." As it turned out, the routine ran out before the music ended. Sensing this, Dean instinctively picked up Ginger and twirled her around, surprising the hell out of her. And, as usual, it looked to America as though it had all been planned that way.

A new young singer named Linda Ronstadt was scheduled to be on one week, and since I knew very little about her I put together a medley that might have been titled "Dean and Fill-In-Your-Name." Anybody could have done it.

Linda arrived for her scheduled meeting in the rehearsal hall right on time. Good. We were to discuss her solo, anything she chose, and rehearse the duet with Dean. Good. Ordinarily, first-time guests spent a genial moment in Greg's office before coming to me, but this time Greg knew less about Linda than I did and suggested I begin the niceties. Good.

In walked a short young girl in a T-shirt and jeans, a stray from an NBC tour perhaps, a tourist from Kankakee maybe. No, this was Linda. Behind her, in contrast, was a battalion of agents and managers, mostly young studs dressed in Madison Avenue uniforms: white shirt, dark suit, and tie. Not good.

Linda was friendly enough, but her entourage made me chuckle. Rather than offering any greetings, they checked the room out, staring around to see how big it was and who was in it. Since they didn't know anyone there, I stepped up and introduced myself and welcomed them to the show. I tried to direct all that to Linda, but a black-suit-and-tie screen went up between us almost immediately. I thought I heard Linda mumble something like "hello," but then she maneuvered over to the piano. Ah, something we can relate to. I decided to be especially gentle.

I pushed my way through the suits and got to the keyboard side of the piano. I told her I was anxious to hear the song she'd selected for her solo. Linda stared at the piano keys. One of her people answered for her. "Her latest record."

"Of course, of course," I agreed, having no idea what that might be. A copy of the sheet music was slapped in front of me. The male hand that put it there was freshly manicured, the nails heavily covered with a natural polish sheen. "I've just come from Beverly Hills," that hand seemed to say.

I turned to Linda and said that I loved her new record, even though I hadn't heard it. I thought diplomacy was important. The boys were pleased there'd be no argument about her solo.

I shoved in closer to Linda and asked her if she'd like to hear what I'd put together for her and Dean. In answer, her eyes darted over to the closest man in her entourage. There was a tap on my shoulder.

Me in a hair-raising sketch with Dean and Peter
Graves

"Could I speak to you in private?" the man asked.

We moved away. "Linda doesn't sing with anybody else. She goes it alone. You dig, man?"

I had to plead my case. "But this is *The Dean Martin Show*. The guests always do *something* with Dean."

"It's not her thing, man," insisted the manager. "She'd not only feel uncomfortable singing with Dean, but it wouldn't be good for her image. You got it?"

Trying to work things out, I suggested that perhaps we could come up with something that *would* be comfortable. I should have known that wasn't going to be easy. I

Greg and me standing in for Peter Graves

Phil
Silvers
and Dean

made several off-the-top-of-my-head suggestions, trying to relate them to Linda and
not the gang. But she found something in
space to stare at, something that seemed
far more interesting than our conversation.

"Don't you understand?" the man continued. "She doesn't want to sing with Dean
Martin. He's old-fashioned." Linda's pre-
occupation with absolutely nothing became
more so.

It was time to get Greg. I excused myself
and made another of those "send in the ar-
tillery" calls to headquarters. I explained
what had happened and fully expected Greg
to pull out another one of his miracles. But
to him, Linda Ronstadt, at least at this point
in her career, meant nothing.

"If she doesn't want to sing with Dean,

that's her problem. Just say good-bye to her. That's it."

He didn't even want to bother to come and meet her. I relayed the message to her manager, who leaned over and whispered something to his client. Then he turned to me. "Thank you very much." The boys seemed relieved, but Linda looked sorry. They walked out without another word and we never heard from them again. It seemed like Linda was at the mercy of these career-starters who were playing the Hollywood game. Greg simply didn't have time for that.

The little scene prompted a new order from Our Leader. Agents and managers would be forbidden to come to rehearsals, sometimes even tapings. They tried to argue with him about it, saying they were only protecting their clients. Greg countered with the fact that he was not only protecting *his* star but the show as well.

"If *we* don't know what's best for everyone by this time, then take it up with our ratings!"

Many producers would be afraid to deal so harshly with important people, but Greg very calmly announced to them all that if they didn't like our rules, they could go somewhere else. Greg ran a tight ship and had the courage to hold fast to his rules. It made my job considerably easier.

It surprised us all that Peter Graves of *Mission: Impossible* (and later *Biography*) fame was so musical. In fact, unlike with most guest actors on the show, we delighted in producing numbers starring Peter with the singers and dancers. He sang surprisingly well, played a mean clarinet, and was an all-around nice guy.

His rounds with Dean were equally enjoyable. Usually focusing on men's pre-politically incorrect superiority to women, the sketches came off as sheer fun and games. The women's lib movement couldn't possibly have taken them seriously.

Every time Phil Silvers came on the show it seemed to be an opportunity to bring out the old burlesque sketches. Phil was around when burlesque was popular, so he knew just about every shtick from every stage in the country. Dean loved doing all the corny bits, and Phil never held back. Everything was full-out. Phil starred on Broadway in the burlesque musical *Top Banana*, and I used that theme song for most of the routines on our show.

He also brought our writers several old

Dean, Phil Silvers, and Mary Ann Case

vaudeville skits that we updated for Dean. He was a dear man and just as funny off-stage as on. And who could ever forget that wonderful, beguiling "glad to see ya" smile from that great trouper.

Joey Bishop, a member of the Rat Pack, paid many visits to Dean's show. Like Phil Silvers and Bob Newhart, Joey brought us skits of his own to adapt to Dean and our TV audience, including routines they probably did together when the Rat Pack played Vegas.

Doing impressions, however, was something Dean wasn't very good at.

"All it takes is an ear," Joey explained. Then he proceeded to show Dean his re-

Dean and Joey Bishop share the stage with three stars of the 1970s: (*top*) Jo Ann Pflug, (*center*) Dick Martin, (*bottom*) Karen Black

markable impressions of Marlon Brando, Bette Davis, and even Charles Laughton.

Dean came back with his best shot, Cary Grant. He would slide into his Cary Grant impression many times on our show for no particular reason, except that he knew he had it down cold.

"Why do you have to be so good?" Joey asked. "You're terrific. Second only to Oral Roberts."

"Yes, I have a certain amount of talent," Dean agreed.

"You know," Joey conceded, "you're a pretty fair singer, a pretty fair actor, and a magnificent drunk!"

When Bob Sidney became our choreographer, he brought Bob Street on board as one of his dancers. Street, as he was always called to differentiate him from Sidney, had two important jobs on our show: he was a terrific dancer and he was a great laugher. I mean, a really great laugher, with a big, loud, "har-har" laugh that was infectious to anyone even remotely near.

Greg insisted that he be close to the stars whenever we did a sketch, to egg them on, as well as the audience. Street never held back, and his laughs were heard well above any audience reaction. Even when Greg was shooting shows that had no dancing in them, he'd hire Street to hang around. Greg knew Bob could make anything funnier than it was by simply adding his wonderful guffaws. Greg considered him a great asset to the show. When Bob Street laughed, the whole street laughed.

Bing Crosby made still another appearance that year, and I decided to pin him down regarding the rumors that he really wasn't born in my hometown of Tacoma after all.

"Well, the record shows I was born in Sequim and moved to Tacoma after only a few days on earth. But what the heck, Tacoma likes to claim me." The city has a Crosby Historical Society and regular salutes are given. His movies are shown; his records are played. "A lot of the facts are inaccurate," he continued, "but it's nice to know so many people cared that much about me."

His wife, Kathy, was never around our studio. There were stories that she didn't like Dean and definitely didn't approve of his booze-and-broads image. Years later, when Dean roasted Bob Hope, the logical guest was Bing. He was invited, but no reply came after several requests.

Charles Nelson Reilly
and Ruth Buzzi

"Bing is hunting in Idaho and can't be reached," was the answer we got from his house.

Finally writer Kendis Rochlen got through to Kathy. "Why won't Bing do our show?" she asked.

"I won't let him. My dear, *you* wrote something really rotten about me and Bing when we first got married and I'll never forget it. This is my way of returning the favor."

It's possible Bing was never told about the Hope roast. One would think he'd be too much of a pro to turn down a show honoring his lifelong partner and friend. Besides, Bing liked Dean.

We gave up. Kathy was too devious.

We had done several specials with Jonathan Winters, and his guest shots on Dean's

Eddie Albert, Dean, and Jonathan Winters

shows were always wonderful. Greg allowed
the comedy genius full freedom, and Dean
was always up for whatever character Jonathan invented. Jonathan still considers
Dean the best straight man he's ever had
(although Dean got a good share of laughs
himself).

My job was trying to find something Jonathan could do in our musical finales. He was
tone-deaf, he said, and didn't care who
knew it (we all knew it). But this year our finales were the lip-sync Dino Vino disk
jockey bits. Jonathan loved them and nobody did them better. He seemed to be able
to capture the right feeling the first time he
heard any record. His mind, of course, took
him into crazy twists and turns. We'd suggest outrageous wigs and costumes to go
along with it, all of which Jonathan made the
most of. He invented so many things far be-

yond what our staff could ever come up with.

Charles Nelson Reilly, like Dom DeLuise and Paul Lynde, was an almost-regular on the show and great with Dean.

"Dean was so supportive of us all," Charlie later recalled. "He would very often turn away from the cue cards and ad-lib, and it was so exciting to go along with him, whatever path he wanted to take. He was just having a good time. In fact, all those years on Dean's show were good times."

When Charlie said a line or did some funny business that scored with the studio audience, Dean would playfully step on Charlie's toes if they were standing, or nudge him if they were seated.

"He was just so genuinely happy that you did well."

Dean wasn't asking people to "keep those cards and letters coming in" anymore. That was one of the season's changes. But during that seventh season, some of the old show, and some of the old Dean, was creeping back. The girls' dresses got skimpier, jokes about sex were allowed back in the sketches, and Dean could occasionally hold that glass of whatever.

Dean and Jonathan Winters

By the end of the year Dean closed the show at his old stool, with a well-endowed, bikini-clad young lady handing him a drink. There was a swizzle stick in the glass that bore a short rhyme that he read to the audience:

I've been doin' this show for seven years.
It gives my heart a little tug.
Thanks for letting me into your living rooms,
I hope I didn't spill anything on the rug.

Those lines closed our show that season, a year that began as pure and proper as we could possibly get, but whaddaya know! We were back to booze and broads!

Dean, Bing Crosby, and the Dingalings

10 The Eighth Year – (1972–1973)

A Link With a Lion

Our gradual change from purity to pure sin in the previous season caused the network and its affiliates to point the finger of shame at us even before our eighth year began.

As Kay Gardella in the *New York Daily News* put it:

When it comes to television morality we guess we're about as big a prude as you can find. We're as sensitive as the next person to unfavorable material inundating the home viewer. But we ask you, did the NBC censors have to declare war on our favorite television star, Dean Martin? What's wrong with an occasional double entendre in the 10 p.m. Thursday night NBC hour?

Plenty, apparently, according to NBC affiliates and Martin critics who have been bombarding the network with letters demanding a clean-up campaign on The

Dean Martin Show for next season. And by the looks of things, they're going to get it. Producer-director Greg Garrison, a pretty hip guy, has promised that affiliates' fears will be allayed when fall production begins.

But what will that do to Dean? Martin's image has always been that of a loveable rogue who likes to occasionally imbibe and who has more than a passing eye for women. For our money, the handsome, graying singer has successfully played the bon vivant, and, despite his on-camera shenanigans, never loses his charm.

Greg decided to make changes, all right. The show was almost completely revamped. "We paid our dues and all bets are off," he said. "We were living in a monastery for almost a year. The changes last season were my fault. I was the guy who took the shot, not Dean."

Fortunately the NBC Standards and Practices fellows hadn't stopped Greg from emphasizing pulchritude, so the Dingalings were back. He also added a bikini-clad "cue card girl" for the station breaks, with Dean reading the words written "all over her body." Rodney Dangerfield was hired to host a "Dean's Place" nightclub segment that gave the Dingalings a chance to sing a

Dean and Rodney Dangerfield

30-second "perfectly awful" production number. That brought on Rodney, who did a short "nobody gives me any respect" monologue before Dean entered for some by-play between the two of them. That part was always taped before Rodney's mono so that Dean wouldn't have to hang around any longer than necessary.

Women's libbers again took offense to a new part of the show in which a guest shows up with his or her favorite pet animal. Lloyd Bridges, for instance, would walk in with a St. Bernard, Fess Parker with a camel, someone else with a kangaroo. In each case Dean had the same pet, a voluptuous girl in a tiger costume on a leash.

Dean to Glenn Ford: "I understand reindeer can stand very low temperatures."

"Yes, in fact they classify reindeer according to what temperature they can stand. My pet is 42 below."

"What a coincidence. Mine is 42 above!"

Right from the beginning, Greg knew we'd get a lot of mileage out of that bit. We did. But from unexpected areas. The full wrath of the ERA over the tiger, the Dingalings, the Golddiggers, and Greg's chauvinist attitude prompted a big picket line in front of our offices with signs like "Put Clothes on the Dingalings or Send Them Back to the Zoo!" and "Greg Garrison Hates All Women!" Once again, their protests made the local news and gave us lots of good publicity.

Of course, nothing changed. Except the mail. Kay Gardella's article in the *New York Daily News* brought a flood of pro and con letters which she shared with her readers:

I have ten grandchildren and we all love The Dean Martin Show. *The Dingalings are so delightful and what's wrong with the pole and the closet, especially when the show goes off at 11 p.m. I feel so good for days after.*

— *The Ratto Family*

Your defense of The Dean Martin Show *was as nauseating a column as I've ever read. Why you would use hundreds of words trying to sell such a cheap, bawdy production*

is indicative of your sense of values.
                            — A TV Viewer

I am with you. I love The Dean Martin
Show, *too! When I read your column today
I immediately wrote to NBC to let them
know there are people who enjoy Dean as he
is. I also told them I'd rather have my three
children watch a Dean Martin "put-on"
than* All in the Family.
                            — Anne M. Frey

*I read your column and I am in complete
agreement with the critics of the Martin
Show. I'm surprised that you condone such
carrying-on.*
                            — A Weary Viewer

*I am so angry to know that so many people
are resenting Dean Martin's Thursday
night show. I just wait for Thursday night to
roll around for his program. Don't people
ever resent some of those talk shows which
discuss everything from abortions, people on
dope, living together without marriage, and
so on?*
                            — Mrs. A. D. Biase

NBC's mail department was inundated
with the same kinds of letters, split about

Ed Kerrigan, Dom DeLuise, and Kay Medford

evenly down the middle, all of which told Greg and Dean that a hell of a lot of people out there were still watching us!

Here we were in our eighth year and whatever the show looked like, I was going to have to come up with something new to disguise the fact that Dean only knew a certain number of songs and wouldn't rehearse.

"How can we come up with new twists?" the writers asked me. "Is there anything we haven't done?"

One of the things we desperately needed was a new finale. It seemed we'd exhausted all those, too. We'd tried burlesque blackouts, imitations, and record lip-syncs . . . We needed something fresh.

And we got it. It was fresh *and* exciting. Greg made a deal with Jim Aubrey, who had

Dean, Nipsey Russell, Dom DeLuise, and Peter Sellers

taken over at Metro-Goldwyn-Mayer, to use clips from any of their movies released before 1955, those from what were called the "Golden Years" of musicals.

But what would we do with them? How far could we go? There were so many gems to choose from. MGM had made the best that Hollywood has ever known. There was song after song. We decided to take one movie a week, preferably those with the greatest number of hit songs, do sections of songs with our cast, and then dissolve into the film clip.

My first selected movie was *Gigi*. Plenty of good stuff in that one. Ed Kerrigan, fresh from a syndicated series with *The Golddiggers*, was Dean's choreographer that year. One night each week Ed and I would set up viewings of two MGM musicals. We'd rush over to the MGM lot with our chocolate shakes and Big Macs and sit there in the same screening room that Louis B. Mayer and all his big stars had used to watch the daily rushes. Ah, the ghosts lurking about!

We made notes, recorded every musical number on a little Sony cassette, and often asked the projectionist to replay sections so we'd be absolutely sure of what we wanted to use. Then we'd go back to the office and put it all together. It was important to make sure our musical keys for the "live" sections would blend into what was on film. Geoff Clarkson and Van Alexander transcribed the brilliant MGM scores almost exactly. There was a smooth transition from, say, Dean singing "The Girl Next Door" to Judy Garland's "Boy Next Door."

Dean and
Gene Kelly

There were no restrictions. We used as much film as we wanted, usually the strongest and best segments.

Since Gene Kelly was to be on our first show to air, we picked his Academy Award-winning *An American in Paris* for that week's finale. Because the MGM production numbers were so overwhelming and beyond what we could put together at NBC, it was obvious that we would sing *into* the clips, saving the best for the last. There'd be half a chorus of a song with Dean or Gene, followed in smooth order by the MGM clip.

We had put seven shows on tape before the Gene Kelly one aired, and still no re-

Nipsey Russell, Dean, Gene Kelly, Dom DeLuise, and the Dingalings

strictions were placed on Ed or me. But what did we have to give in return for this amazing gift? No money was involved. All Aubrey wanted was a sixty-second plug for the studio's latest movie at the end of our show. It seemed fair enough to us after all that we were getting free, but NBC said they couldn't "give away" that much time. And if MGM couldn't get the plug, we wouldn't get the film clips. There went ten minutes from every show.

Negotiations went on right up to the day the show aired. It was finally agreed to announce at the end of the show that "clips from *An American in Paris* courtesy of Metro-Goldwyn-Mayer, whose latest release is *They Shoot Horses, Don't They?*" or such-and-such. That agreement was resolved about an hour before Thursday night's premiere of *The Dean Martin Show.*

Everyone seemed happy, especially Ed and me, for we were able to freely lift incredible scenes from *The Band Wagon, Three Little Words, Singin' in the Rain, Meet Me in St. Louis,* the best of the best. We carefully worked them into musical sections with Dean and his guests, as well as our company of players: Dom DeLuise, Nipsey Russell, Kay Medford, Lou Jacobi, Marion Mercer, Tom Bosley, and the Dingalings. Everything

Dean, Lynn Latham
of the Dingalings,
and Dom DeLuise

was a joy until our tenth week. I began to get calls from Jack Haley Jr. over at MGM.

"Please don't use the 'Babbit and the Bromide' number from *Ziegfeld Follies*." Very polite.

I knew that it was the only number Fred Astaire and Gene Kelly ever danced together and I definitely planned to use it. After all, Mr. Aubrey said there were no restrictions.

"I know," Mr. Haley insisted, "but we have a little project in mind over here that we might want to use it in."

Greg told me to ignore Haley, that we had made a deal with Aubrey and he was expected to stick to it. But Haley kept calling back, asking us not to use some clip. I was as polite as he was, but I used the segments anyway.

Apparently Haley never monitored our shows because he never called back to ask *why* we used a clip he had asked us not to. Aubrey did send me a memo that included a few films we were definitely *not* to use because of contractual problems. Two of them were big Irving Berlin musicals, *Annie Get Your Gun* and *Easter Parade*. I assumed Mr. Berlin had them locked up somehow.

One day I got a call from the Berlin office. A nice man asked me if there was any particular reason we weren't using these films in our finales. I informed him of the memo from MGM. He asked if I would be interested in those two Berlin classics.

"Are you kidding? They're terrific movies."

"I'll call you back."

Two hours later the nice man called me again with the remarkable statement that "Mr. Berlin has personally called MGM and made it possible for your show to use his two pictures in your finales. He would like you to use as much of the music as you like. Mr. Berlin would also like you to give his very best to Dean and to congratulate him on his fine show."

When I threw that at Dean he told me that Irving had long planned a super production at MGM called *Say It with Music*, using a great number of Berlin songs. His first choice for the lead was Dean and still was. Dean was anxious to do it, but apparently the many upheavals at MGM kept putting the project on the shelf for lack of funds and it was still sitting there.

If Jim Aubrey was trying to keep certain clips from us, Irving was having none of it. We were proud and happy to be able to use his two pictures, which, by the way, he refused Aubrey permission to use in his *That's Entertainment*, the secret project he was working on while we were doing our own version of it.

We didn't want to make our live sections of the finales anything like the clips, so Ed Kerrigan and I had to come up with new ideas for each song. Going into Astaire's

Me and Van Alexander, our arranger

"Drum Crazy" number from *Easter Parade*, for example, Ed had Dom DeLuise singing the first half of the chorus while pounding on two huge white kettle drums. One of them was filled with milk and when Dom finally hit it with his mallet half his arm disappeared in the drum. Milking the applause, as it were, he eventually jumped in and swam around in the milk. Then we cut to Fred with drumsticks, tapping around a toy store.

Although the writers will probably disagree, I thought the finales were easily the best part of the show that season, an opinion verified by the fact that our ratings went up during the last half hour. People obviously tuned in to catch the finales.

As happy as Ed and I were with the finales, though, it was tough to get the cast excited about them. From the very first Martin shows, finales were something they had to put up with, bits of fluff that most performers felt needed little or no effort. Their solos came first.

Gene Kelly, fortunately, gave these finales considerable attention. He had so many revealing stories about the making of *An American in Paris* that at one point we considered using them as part of the medley. But we didn't want them to look like documentaries.

As for the others, only Dom DeLuise gave the finales some thought and energy. Whenever he entered any room for any reason, Dom always pulled the rest of us up. Because of his attitude, Kerrigan and I used him more than the others in the finales. We knew he'd come through. He was particularly helpful in keeping Dean on his toes.

When we did *Till the Clouds Roll By*, I used a clip of Frank Sinatra singing "Old Man River," and let Dean sing the first half of the song. It was one of the few times that Dean *asked* to do something over. As a matter of fact, he did it three times.

"I wanna show that son-of-a-bitch I can sing that song better than he can!"

Gene Kelly and the Golddiggers

It was my first crack at editing, something I enjoyed very much. I was determined to make those musical edits smooth, cutting the tape on the beat of the music, as if our guest star was continuing on with the song. I also had to make sure that the transition from film to live was in the same or a relative key. The cutting back and forth in those finales was hailed by all, including MGM head Arthur Freed, songmaster Irving Berlin, and especially Greg.

Two other new segments of the show had mixed results. Nipsey and Dom played barbers in an NBC barbershop, a place where Dean and that week's guest could drop by. It was loaded with the usual macho talk that flows in barbershops, most of it on cue cards, but a lot of it ad-lib.

Dom and Nipsey were great at that. Dean was no slouch, either. Peter Sellers was one guest who found this particular segment difficult. He explained that he liked to be a character, and in the barbershop we were asking him to be Peter Sellers.

"I don't know how to be myself," he said. "I'm dull."

It bothered him so much that just as we started to tape the barbershop segment he invented a character for himself. He didn't

Dean
and Jack
Benny

change any lines, just read them as a very flamboyant gay, completely surprising and delighting Dom and Nipsey. The ever-present censors were confused. They scanned through their scripts and saw the lines were all there as written. But Peter's new interpretation gave them new meanings. Dean hadn't entered yet but was breaking up off-camera. When it was time for his entrance, Dean decided to be a little swishy himself.

"Hi guys!"

To which Nipsey ad-libbed, "I'm beginning to think I'm the token *straight* in this sketch."

Like all our actor guests, Peter liked to re-

Dean and
Peter Sellers

hearse a lot. Still, without changing any lines he fell into the utter madness of this new character that he had created, easily making it *his* sketch.

During the week, we all thought Peter was a little weird. He would spend a good deal of time standing on his head over in a corner of the rehearsal hall. When we got to the *Meet Me in St. Louis* finale and one of his songs was "Skip to My Lou," he insisted on speaking the square dance calls with an extremely hard American "R," and it came off as condescending. However, he relaxed and thoroughly enjoyed dueting with Tara Leigh on "Have Yourself a Merry Little Christmas."

Our sitcom experiment was still around that season, only now it had moved from Kay Medford's living room to "Gus's Diner." Kay's and Abe Vigoda's characters were married and the diner was their new place of business, where any show guest could drop in for coffee and laughs. Tom Bosley and Marion Mercer were let go, and the laughs grew thinner and thinner. The whole thing petered out midway through the season and was dropped. Our sitcom had flopped.

There was another new look, however. Dean's face. He was never one to use much

The Dingalings

makeup, if any, but we all noticed some cosmetic work had been done. Probably nothing more than tightening a few lines here and there, we figured.

Well, we knew who was behind that. The hair kept changing, too. Every time Dean did a song with the Dingalings or the Golddiggers, Kathy Hawn was on the stage like a fox, watching every move. Her snide remarks were overheard by everyone in the immediate vicinity, certainly the girls and crew, probably some of the audience, and Dean.

Her outspoken, salty evaluation of everything we were doing out there on Stage 4 was naturally the topic of conversation of the staff and crew after each taping.

"Is this little broad going to make things difficult for us?" we wondered. So far there were no complaints from Dean's dressing room, not even by the roundabout Dean-to-Mort-to-Greg route. (Dean would never complain himself. What few negative views he might have always came from friends.)

The girls on the show and the lady guests were performing with Dean, not trying to take him away from Kathy, but she must have thought differently. It began to shake us a bit, so the Dings and the Golddiggers were asked to be careful.

Suddenly, a wedding was announced, but none of us received an invitation.

Beautiful opera star Anna Moffo was a guest and made it known that she wanted to be "one of the guys." I did a little "Today and Yesterday" medley for her, Lloyd Bridges, and Dean with some pop songs in old styles. It went along well enough until the very last note. The three of them were supposed to pose with Anna on Dean's knee and Lloyd around the back. But Anna somehow missed the knee and fell over Dean, dragging Lloyd along with her. In spite of a very short skirt she managed to keep her dignity. But when Dean got up he discovered he'd split the back of his tux pants, revealing some loud undershorts. The camera zoomed in and followed him offstage, past the crew and into his dressing

Dean, Anna Moffo, and Lloyd Bridges

room, an unplanned but hilarious finish to an uninteresting musical bit. It was all left on tape and to this day people remember it, saying, "What a great medley that was!" Thanks, Dean.

Hugh O'Brien and Monty Hall joined Dean and the Dingalings for a ladies medley and managed to prove conclusively that their TV roles suited them better than singing and dancing. William Conrad, on the other hand, showed off a remarkably strong baritone in his solo "The First Time Ever I Saw Your Face." In his medley with Dean he often enjoyed the sound of his voice so much that he'd refuse to get off a note, extending the playing time considerably.

Dennis Hopper, no singer, and Charley Pride, a good one, were teamed in a section of the *Singin' in the Rain* finale. They both tried hard to make "Good Morning" work, but it was not to be. Choral Director Jack Halloran and I had to dub in their voices. The main trouble was that they had no sense of rhythm. Even walking in time with the music was a problem and when we got to the syncopation of "Broadway Rhythm" with the rest of the cast, I was forced to put all those nice off-beat notes *on* the beat to keep everybody together.

Nipsey kidded Charley about it: "Just proves we ain't *all* got rhythm!"

Danny Thomas was a funny man in the sketches and fine in the finales, but difficult during rehearsals. He was one of those guests who resented the fact that Dean wasn't around all week while *he* had to rehearse. He more or less went through the motions, never giving us a clue as to what he would be like on-camera.

It was hard to get him to seriously rehearse anything. He much preferred spending his time telling disgustingly dirty stories that made us all cringe. No prudes in the house, of course, but Danny didn't care who heard his decidedly blue humor. The ladies in the room walked away and even the raunchiest men barely snickered. It didn't stop Danny, though. He just kept pouring out the filth until he emptied the room.

Dean,
Charo,
and Danny
Thomas

The Dingalings were particularly gorgeous that year. Greg allowed Bob Fletcher to dress them to the teeth, although he occasionally yelled at Fletch that "the whole week's budget goes into costumes for those broads!" Tara Leigh and Lynn Latham remained, and the new girls were Jayne Kennedy, the most stunning African-American girl I'd ever seen, and Helen Funai, a beautiful Asian girl who had been one of the background dancers on our first shows. We made the attempt to upgrade their music and make it more contemporary. But their voices were definitely not Top 40, what with

Tara's high soprano, Lynn's Julie London softness, Jayne's borderline passable contralto, and Helen's inability to find any of the notes. Melissa Stafford was on duty to sing for Helen in all the prerecords. It was even necessary to prerecord their medleys with Dean so there'd be no embarrassing moments. Dean, of course, sang live.

It was definitely touch and go musically, but Greg argued, "Who's listening to them sing? Faces, faces! That's what's important." And they were allowed to touch Dean again that year.

Most of the girls on the show through the years found Dean very sexy and some thought of him as a father figure. They all liked him. He charmed them all and wanting to touch him was a normal reaction.

Jack Benny made one of the last professional appearances of his life with us that season. Unlike Danny Thomas, he enjoyed rehearsing and made no fuss about Dean's absence. He seemed shaky that week and we all tried to be especially nice to him. It was easy. He was a kind and gentle man.

This was the year that saw Dean descend from above the stage in a chaise lounge elevator after each Dingaling number. Jack

Jimmy Stewart, Dean, and Frank Sinatra Jr.

thought it would be great fun if he appeared there instead of Dean for the Dings' medley.

We did him one better. We hired four senior ladies to be Jack's Dingalings. They met him when he landed, petting and pawing him as our girls did Dean.

Dean brought in his Dings and the scene became a challenge, with Jack setting out to prove that he could be just as much a sex symbol as Dean. His delightful "golden girls" screamed like adoring fans at everything Jack said.

"They get excited when they're away from the home," he explained.

The barbershop sketch — Dean, Dom DeLuise, Nipsey Russell, and Mike Connors

"Be careful, Jack," Dean whispered. "They'll hear you."

"Believe me, Dean, they *can't* hear us!"

Although quite physically infirm, Jack was still at his peak in sketches and even while doing "Johnny One Note" in the finale (he took out his violin and screeched out one sad note).

The MGM finales became longer as everybody got as excited about them as I was. It would be impossible to do all that today. We got away with murder. The studio's famous library was gobbled up. New restrictions and revised union contracts later made it impossible to use such long hunks of those great performances. But we sure had a great run with them!

Besides the fun of the finales, we found ourselves enjoying more freedom in associating Dean once again with booze and broads. Women's lib groups continued to voice objections, but even the network ignored them. An outspoken advocate of antichauvinism, Helen Reddy made it a point to publicly declare Dean and our show "disgusting" and "demoralizing." A lot of today's world might see it that way, too, perhaps, but we thought of it as nothing more than doing what comes naturally. And we had no intention of changing Dean's image again!

Dean and Orson Welles

11 The Ninth Year – (1973–1974)

Clutching at Straws

After eight lovely years dwelling in the top half of the Nielsen ratings, we were faced with a decline over the last few weeks of year eight. There was heavy concern about how we could recapture the lost audience. Was the public getting tired of variety shows? Was Dean's carefree style wearing thin?

We had tried to make each season look a little different, while still keeping Dean's casualness. This coming year was really going to have to knock 'em dead or America was going to turn off.

Twentieth Century-Fox and Warner Brothers had expressed some interest in letting us pick up where we left off with the MGM finales. But after looking over their catalogues, we decided their offerings were second-best all the way. "Besides, we've done that," Greg kept saying.

Ed Kerrigan and I hit on the idea of

"toasting" a different star each week, sort of a musical *This Is Your Life*. It would be a salute to a guest, reliving their lives, showing some film clips, singing their songs.

Petula Clark was set for our first taping and we figured she'd be easy. We could do all her record hits live, show scenes from *Finian's Rainbow* and *Goodbye, Mr. Chips*, and bring on some special friends. But that was about to change.

Greg liked the word "toast," but it immediately reminded him of "roast" and he thought why not? "We'll do an old-fashioned Friar's Roast, poke fun at somebody famous. It doesn't have to be musical. Better if it's not. We'll make it a big part of the show, maybe fifteen or twenty minutes."

Dean, Loretta Lynn, and the Statler Brothers

Bob Fletcher was called on to build a dais on our stage with an auditorium look; we even hired extras to sit at tables in front. Greg's fifteen or twenty minutes grew to half an hour, and as the season went on, sometimes the full sixty minutes. But in the beginning it was just a section of our regular show.

Our first two candidates to roast didn't strike anyone as being exciting enough to draw new viewers. Henry Frankel, our booker, said he was having trouble getting anyone to take part in something that might embarrass them.

Two personalities who said they didn't embarrass easily agreed to be our first victims, Hugh Hefner and Ed McMahon. Roasting Hefner, if done today, might be just naughty enough to be fun. But in the early seventies, the network made sure we kept it clean. Even Hugh looked bored with it all. Just before taping he bounced on stage in a bunny suit, insisting it would be funny to wear it on the dais. When it finally sunk in that he wasn't kidding, there was a half-hour wait while we talked him into wearing a normal tuxedo. Actually, the bunny suit might have helped. It was a boring roast. But it was our first time out.

The other part of the show was routine

Dean Martin programming, except that another new segment was inaugurated. Our summer show that year, *Dean Martin's Music Country*, had done well in the ratings so we figured we could continue with country music in the fall show. The section had its own set, something country music stars hated from the start — Hollywood's version of country, complete with a barn, hay wagon, front porch, dog, and girls in checkered dresses.

Dean had been hooked on country music for some time and agreed to get rid of his tuxedo for this part of the show. The country stars loved the exposure, even if they despised the set. And they all worked well with Dean.

Some of the ladies were so enamored of him they were nervous wrecks. Loretta Lynn, particularly, wilted at Dean's every word. "Oh my God, it's him!" she'd say and forget her musical entrance.

Although we booked a couple of country names each week, the rest of our cast was thrown into the Music Country medleys, often incongruously so. City singers like Petula Clark, Dionne Warwick, the Golddiggers, and even Gene Kelly and Donald O'Connor all donned gingham or jeans. But none of them treated it as corny. Les added

Dean and
Foster
Brooks

a couple of guitar players to his band and
Van Alexander made the orchestrations as
"Nashville" as possible.

Getting Kris Kristofferson to be a part of
the segment was a real coup. He had become a big star, country and beyond, and
Dean was a fan. Kris had one demand, however. He wanted his wife, Rita Coolidge, to
appear with him.

"Who's she?" Greg kept asking, thinking
America would be wondering the same
thing. When I explained that Rita was a
growing country name and had done some
albums with Kris, Greg was not impressed.

"Nobody'll know who the hell she is!"

Henry Frankel did his best to get Kris
without Rita, but it was the two of them or
nobody.

"Okay then," Greg said, " but don't give
her anything to sing. Just let her sit there."

"We can't do that," I kept telling him. "The lady's well-known."

"Not to me, she ain't!"

He finally agreed to let her do eight bars with Kris before disappearing into the farmhouse. It turned out that Kris was very anxious to do a duet with Dean and didn't seem to care what happened to his wife.

"Shi-i-i-it!" he'd exclaim in wonder, every time Dean opened his mouth to sing. He couldn't help himself. He had put Dean on a pedestal. I threw in two of Kris's own songs for Dean, which Dean had just recorded. Kris was even more in awe.

Rita sat around quietly, accepting her lowly position. "See? She's dull," was Greg's reaction to that. "Why wouldn't she be?" I answered. "She hasn't anything to do." Greg eventually backed down and brought Rita back for the final song with Dean and Kris. "Everybody must have thought she was one of the Golddiggers," Greg mumbled as the medley ended.

John Wayne was scheduled to be our third roastee, and probably would have aired as our first, but at the last minute he cancelled, saying he never agreed to it in the first place.

Greg was furious with Janet Tighe and writer Kendis Rochlen, who had set it all up,

Dean roasting Bette Davis

for letting him squirm out of it. They all tried hard, going back and forth with Wayne, reminding him that material had been written and guests were lined up especially for him.

But something was going on in Wayne's private life. He was divorcing his wife and there were rumors that it was because of his affair with his secretary, Pat Stacey. Apparently he was afraid someone might mention that on the roast. It could get out of control and tarnish his image. Old reliable John Wayne? Hadn't he survived many film duds and still remained America's top star? Hadn't he freely expressed his political views without any negative reaction at the box office? Wasn't he the symbol of

Americana that nothing could touch?

Obviously he was vulnerable at that moment. He refused to discuss it any further and was replaced at the last minute by William Conrad, then a pretty big figure in TV's *Cannon*.

Kendis was also a good friend of Bette Davis, who agreed to being roasted if Kendis wrote all the material for her rebuttal. She also insisted that no one come to her dressing room but Kendis, with the exception of Greg, who after all *was* the producer. Bette was probably sorry she'd committed herself but was professional enough to go along with it.

Dean and the Golddiggers in our Music Country segment

Donald O'Connor

Greg gave her considerable confidence, explaining that this was all in fun and begging her to relax and enjoy. His usual persuasiveness did wonders and they seemed to get along famously in those dressing room tête-à-têtes. We had roasted Johnny Carson the previous week and Greg even maneuvered Bette into taping a guest spot to be edited into the Carson roast.

"He doesn't know it, this Mr. Garrison, but I've got him figured out pretty well," she confided to Kendis. "He got another shot out of me for nothing."

When Greg made no attempt whatsoever to hide the fact that he was doing exactly that, she started to laugh. It got her into a good mood for the show.

"The guy's got guts. I like that."

407

★ ★ ★

The Carson roast was the first one to go an hour and the first to have some major guests on the dais. Johnny had personally asked for Joan Rivers because she had guest-hosted so many of his shows. Joan, admitting that Carson gave her her first big boost, said she'd "go through fire for Johnny, but. . . ." And it was a big "but."

She refused to do the show because of Greg. It seems she had auditioned for him several years earlier and thought she had done very well but she didn't get the job. She demanded to know why and Greg told her. "You're too New York and you're too Jewish."

She swore she'd get back at him one day, and she found the day.

Greg didn't want to turn down Carson's request for Joan, so he kept the bookers working on her. She stuck to her guns, in spite of her loyalty to Johnny, and refused to be part of the roast.

Now it was Greg who wanted to know *why*, and when he was told Joan's reasons, he had a predictable reply:

"She's still too New York and she's still too Jewish!"

Mark Spitz fell into the category of a "new

and interesting personality" and was booked on a few roasts. He had just won his seven Olympic medals and was as famous as any Hollywood star. But although he'd been seen laughing on several roasts, his spots at the microphone were always cut out. Greg felt his attempts at written jokes would prove embarrassing to him. Greg tried keeping him over at the end of the tapings to redo his short bits, but all anyone ever saw on the final shows was Mark sitting at the dais laughing.

Another personality still capable of giving us some mileage was Zsa Zsa Gabor. We decided to roast her. She immediately objected because she wanted to see in advance what others on the dais would be saying about her.

Greg tried to explain that that would take away all the fun, but she didn't think so. She was sure everybody would be unkind. Some of them were, but Zsa Zsa didn't understand what they were saying, anyway. Her rebuttal was a disaster because she stopped after every few words. "I can't see the cue cards!" she kept yelling. Curiously, we had spent considerable time before the taping to double-check the distance between her and the cards so this couldn't happen. She even wanted to make the size of the letters larger so she'd be sure to see them. She bungled through her lines to very little audience reaction, so we kept her there after everyone else left and redid her rebuttal entirely.

Zsa Zsa was a guest on another roast and assigned to sit next to Milton Berle and his cigar, both of which she regarded with extreme repugnance. She also fought bitterly over her position at the far end of the dais. She was still arguing about it when the show started. When her name was announced at the opening, Janet Tighe had to literally shove her out on stage, a gesture that overshot Zsa Zsa past her assigned seat, making it necessary for her to ungracefully work herself back to the only seat left. "She's still worth it!" Greg would say. "People like to make fun of her and she's able to take it."

Carroll O'Connor wanted to be roasted as himself and not Archie Bunker. "But who's Carroll O'Connor?" the writers asked. "And what can we say about him?" We compromised. A little of one, more of the other.

As usual with roastees, he was asked if there were any special friends or relatives he'd like on the dais. Carroll suggested his wife and added that she was a good singer. Up to this point there had been nothing musical in the roasts, strictly comedy. But Carroll pressed Greg, and I was tapped to draw up something for her.

I wrote her a special material number about what Carroll had first complained about, "Am I Married to Carroll O'Connor or Archie Bunker?" His wife, Nancy, liked it and was sure Carroll would see the humor in it. She learned it and performed it live near the end of the roast. She did a good job on it, but the show was too long and Nancy was the first to get cut in the editing room. Those other big names on the dais couldn't be sacrificed.

Carroll was not happy that his wife never even appeared on the show. He was one of the few people to call Greg afterward and say he definitely did *not* have a good time.

A perfect person to roast seemed to be Ralph Nader. Although he didn't ask to see what the others would say about him, he spent considerable time with the writers on his rebuttal. He wanted to make sure he got all his points across, including a comment about Dean's smoking, and told him so backstage, much to Dean's annoyance. Greg had to step in and calm down Nader.

"Not tonight, Ralph," Greg told Nader in

subdued tones, so that the rest of the guests couldn't hear.

Who but Greg would think it "fun and interesting" to roast Truman Capote? Surprisingly, Truman thought the same thing. He didn't care who the guests were, except that it would be nice to have Joseph Wambaugh. Or maybe Jean Simmons. We got both of them.

Before we began, Truman was terribly intrigued by all that was going on around him, including the preparations for the roast, and he insisted on watching us tape the other part of the show, the song and dance stuff with Donald O'Connor and Audrey Meadows. "Why don't you put me into one

Dean with Mrs. and Mr. Carroll O'Connor

of those songs or sketches?" he asked me. I wish now we had. What a kick to see Truman doing a soft-shoe with Donald and Dean!

Rich Little did a devastating imitation of Truman in his nightclub act, which Truman said he couldn't wait to hear. He was well aware of Rich's takeoff on him in other shows and considered him brilliant. Rich, too, thought Truman would have made a great performer.

"He could have done an interesting Hamlet," Rich acknowledged. "Can't you just hear him say, 'To be or not to be — that's always been my question'?"

Truman enjoyed the round of slams at his character and his accomplishments. "You all made me feel at home," he said, in his sarcastic way. "My home is in Palm Springs, and nothing exciting ever happens there, either."

Besides Johnny Carson and Carroll O'Connor, full-hour roasts were taped with Don Rickles and Senator Hubert Humphrey. Relegated to thirty minutes or less were sport jocks Joe Namath, Hank Aaron, Bobby Riggs, Leo Durocher, and Wilt Chamberlain. Also in other half-hour versions were Redd Foxx, Monty Hall, Kirk Douglas, William Conrad, Rowan and Martin, Senator Barry Goldwater, Jack

Klugman and Tony Randall as a team, and then-governor Ronald Reagan. Reagan, a good friend of Dean's, was at the peak of his fast-talker ability, spitting out a stinging and hilarious rebuttal while at the same time thoroughly enjoying the barbs against him. On hand to throw insults his way were Phyllis Diller, Jack Benny, Jonathan Winters, Nipsey Russell, Dom DeLuise, Don Rickles, Audrey Meadows, Pat Henry, and in his usual voiceless capacity, Mark Spitz.

Reagan did guest spots on several other roasts, mostly because he liked Dean and the show. There were others who said he liked the free publicity. It didn't hurt and it made him very human. He often said he wished he had a show like Dean's to play with, that he could be just as cool and relaxed. "The only difference between Dean and me is that although we both made movies, I knew when to quit."

It was a short half hour and funny enough to be edited into our premiere show of the season, certainly more acceptable than Hugh Hefner or Ed MacMahon.

The roast experiment worked. It worked so well, in fact, that NBC asked for roasts *only* from now on. No more variety, no more singing and dancing, and sadly no more *The Dean Martin Show*.

12 The Roasts:

A Whole New Ball Game

Our tenth season wasn't a series at all. The variety show was kicked out in favor of roast specials, maybe one a month. But NBC wanted approval of who would be roasted.

It took us awhile to get over the fact that the music was taken out of our musical comedy shows. We thought our version of the variety show was different — because of Dean. But the end was near for the genre. Only Carol Burnett over at CBS was hanging on.

"Is it us? Have we lost our touch?" We shook our heads and wrongly blamed Dean himself for our downfall. *He* had changed, not us. It was true that he was not as bright, casual, and upbeat as he was in our great 1965 beginning. Some of the staff blamed Kathy Hawn, saying she had taken the zing out of Dean. We saw the changes. In the old days every first take was terrific. Now we were doing things over and over. He was getting forgetful. And the ad-libs were becoming mean.

We'd heard all the excuses: the ulcers, the breakup with Jeanne, the reported trouble with Kathy, the problems with the kids. All that together, however, wouldn't have fazed a professional man like Dean.

We couldn't face the fact that the public was tired of variety entertainment. We'd saturated the TV screens and now it was time to go on to something else.

Since I had done the editing on our MGM finales, Greg insisted I perform that job on the roasts. He made me associate producer and, the following year, the producer.

Me with Henry Kissinger for the Bob Hope roast

The roasts were simple enough to do, but outside of the fact that they corralled a bunch of big stars together at one time, they seemed sorry replacements for the sheer joy of music, dancing, and eye-popping production numbers.

As for Dean, he didn't really care if he worked at all. He wasn't interested in who we roasted, who was on the dais, or when we wanted to tape. We knew he'd show up because he probably had nothing else to do. He faded into the background, becoming just a figurehead whose name we simply used over the title.

Our first roast of the 1974–1975 season saluted Bob Hope, and the dais was loaded with top names. Since Bob was close to every president since FDR, writer Kendis Rochlen used her connections to see what we could do about getting Gerald Ford, who was in the White House at the time, and Henry Kissinger to do spots. Their respective staffs agreed, so I flew to Washington, D.C., to make arrangements to tape some one-liners our writers had concocted for them, explaining that we'd try to catch them whenever they were free. Ford agreed to do it, but Kissinger, working hard on a Mideast agreement at the time, didn't seem anxious to allow his life to look the least bit frivolous.

But Kendis kept trying, and even as I arrived in the nation's capital there was still a good chance both of them would sit still for five minutes and say a few words about good old Bob Hope. I arranged for the most mobile TV unit NBC had in Washington to stand by for a full day, sticking to our promise to interfere in their affairs only when *they* wanted us to.

The hourly bulletins from Kendis on the Coast kept me updated. "Ford is okay, Kissinger says no." "Ford may be able to work it in, Kissinger says maybe, if he finishes three important meetings." "Ford can't be reached, Kissinger says okay."

My camera unit was still standing by when I told White House go-between Bob Hartmann that although I didn't want to disrupt any important government decisions, I did have a problem. He said that I should be ready to tape either Ford or Kissinger as they were going from one place to another. I hoped that Hartmann would be able to tell me where and when I should be "ready."

At 4 p.m. I got a call back. "Ford had to fly to Mexico City. That's the way things go here. But Henry says he'll do it. He got the material you sent him, says to get your tape people to the eighth floor of the Senate of-

Ruth Buzzi
as "Gladys"

fice building. He'll be passing through there around six and he'll give you, oh, maybe fifteen minutes. Good luck."

Our unit was portable enough to set up anywhere as long as there was AC current and we found exactly the room Hartmann had described. It couldn't have been set better by a Hollywood decorator: a wood-paneled office with a big American flag and a portrait of George Washington. A perfect backdrop for Kissinger. He must have known.

Six o'clock came and we waited. Nothing. 6:30. Maybe he's been held up by some pressing new business. 7 p.m. Maybe he's forgotten. 7:30. Maybe he's changed his

The Bob Hope roast: (*back row*) Rev. Billy Graham, Nipsey Russell, Jimmy Stewart, Rich Little, Johnny Bench, and Mark Spitz; (*second row*) Howard Cosell, Flip Wilson, Zsa Zsa Gabor, Neil Armstrong, Phyllis Diller, and Sugar Ray Robinson; (*front row*) Greg, Ronald Reagan, Dean, Bob, and Gen. Omar Bradley

mind. At a little before 8, a shuffling of feet. Half a dozen burly FBI types burst in, searched around the room, asked some silly questions, looked at my credentials, and announced that Kissinger would be there in ten minutes and hoped we'd be ready for him. "He hasn't much time," they added.

Exactly ten minutes later Kissinger arrived, introduced himself to me, and asked if there would be cue cards. Yes. Could he read them on camera? Yes. Could he change a couple of words? Of course. Could he eliminate a couple of jokes? Well, maybe . . . but Mr. Kissinger, this is supposed to be *funny.*

"Who else will be on the show?" he asked. I told him the many celebrities we had booked and his shoulders drooped a bit. "You know, I feel a little out of place making these jokes on a show with so many polished performers. Maybe I shouldn't try to be funny at all. What do you think?"

I jumped in quickly. "The material is, shall we say, more *humorous* than *funny,* Mr. Kissinger. There are no big guffaws there. It's just nice and pleasant, and I think it might look strange if you were totally serious."

"You're probably right. Maybe I could put in a nice little closing . . . you know, say something about how much America loves Bob."

"Oh, what a nice touch," I told him, "especially coming from you." I figured it didn't matter what Kissinger said; I'd edit it so the best parts came out on the final show, anyway.

"If it's all right with you," he continued, "I'll just make it up when I get to that spot."

I had brought along a makeup girl, just in

case, and to my surprise, he asked for her. I guessed he was used to that sort of thing by now. Anyway, she was young and attractive and the conversation switched from Bob Hope to *her*. Remembering the Secret Service's warning that we only had about fifteen minutes with him, I interrupted the little affair. "Whenever you're ready."

"Oh, I thought you were setting up some lights," Kissinger quickly explained. "Sorry."

With a clear "see ya later" gesture, he left the makeup girl and stepped toward the camera.

"Where do you want me?"

"Between the flag and George Washington."

"Oh God, you showbiz people are so obvious."

The lights shot on. I said, "Action!" in a moderate tone, due to the small formal surroundings, and suddenly his smile vanished. As he began to read the cards, he became very proper and stiff:

"There was some doubt as to whether I could make an appearance on this show because of my tight schedule. But Bob said if I didn't appear, he'd stop paying taxes — and you know what that would mean to government solvency."

He stopped for a minute. "Is that supposed to be funny?"

"The laugh track will think so," I told him.

Before we began I had asked him to wait a beat after each laugh line in order to put in some recorded laughs later, but he waited an incredibly long time, probably overpacifying me. He must have known these were not gems. But still, with Henry Kissinger saying them, who knows?

He finally got to the end. The whole thing was twice as long as it should have been and terribly draggy.

"Uh, could we do it again, Mr. Kissinger?"

"Did I make a mistake? I make so few of them, you know." There was a genuine smile on that one.

I walked over and whispered to him that it would be much better for the show (after all, it *was* a comedy) if he didn't look quite so somber during the punch lines, and if he could please speed it up just a tad.

"Oh sure, sure. I understand." He smiled pretty big this time, but as soon as I said, "Action!" he went back to the frown.

"Oops! We lost a key light," I said in Greg's best fashion. "We'll have to do it again."

Henry began to get a bit testy. "If I give

Dean and Jackie Gleason

you a really big smile this time, can I get out of here?" He knew there was no key light problem.

This time I threw in a big fake laugh of my own when I said "Action!" and Henry looked over at me as though I'd gone mad, probably sorry he got into this in the first place, even for Bob Hope.

He smiled nicely, however, through "Good evening" and well into the first paragraph, then got serious again just as the jokes came up. He was telling me that *this* was the Henry Kissinger he wanted the public to see, and to hell with me, Dean Martin, *and* Bob Hope.

There was no question that this would be his last take, smiles or no smiles, but before he left the room he turned to me with some last words: "I'm sorry, but after a hard day

424

with Congress, nothing seems funny."

After Hope, Greg came up with two more big names and an idea to take the show out of the Burbank studio and into Las Vegas, a logical spot for these sort of shenanigans and a chance to grab some performers working there. He was also able to get some dough out of the MGM Grand for plugging their hotel for an hour on prime time. The whole staff, crew, and a mass of stars got bed and board from the hotel for two to four days.

The regular hotel patrons must have thought the Grand was putting them on. All day long they'd hear the paging system say, "Telephone for Milton Berle," "Angie

Dean, Art Carney, and Jackie Gleason

Dickinson wanted on long distance," or "Call for Mr. Henry Fonda." But they were all there, dining in the Barrymore Room, roaming the gaming tables, or just browsing among the plain folks from Dayton, Ohio.

Our first weekend there, Telly Savalas, an international hero at the time for *Kojak*, was roasted on Saturday, Lucille Ball on Sunday.

Lucy was surprisingly easy, saying she considered it more of an honor than a roast, especially with so many big stars on hand for her. She had personally asked for Henry Fonda and Ginger Rogers, both prominent in her early Hollywood days, as well as costar Vivian Vance and Lucy's husband, Gary Morton. Gale Gordon showed up, too, and Jack Benny, Bob Hope, Phyllis Diller, Ruth Buzzi, Milton Berle, Totie Fields, Rich Little, Foster Brooks, and the king of all put-downers, Don Rickles.

Lucy's rebuttal was accepted by her before we got to Vegas and was never altered. Material was written and okayed well in advance back in Burbank for the roastees and all the guests, except for Rickles, Nipsey Russell, and Jonathan Winters, who were always allowed to ad-lib. But nothing was really settled. Almost all the celebrities wanted changes when we got there — a

Orson Welles, Dean, and Muhammad Ali

word here, a sentence there; sometimes a whole spot had to be rewritten. Our writers worked around the clock to satisfy everybody. Offices were set up at the hotel and a frantic fury of last-minute additions and deletions kept everybody busy.

Tickets for the tapings were free and were given out at 9 a.m. on the morning of the show. The lines formed in the wee hours and the tickets were gone as fast as they could be handed out. Around 2 p.m. the stars would gather backstage for some quick repartee, a few drinks, and a publicity shot of everybody. Dean always arrived just before the picture was taken. He didn't like to share the snappy small talk.

As soon as the barrage of flashbulbs was over, the taping started. It ran straight through without any stops, no matter *what* happened, for about two and a half hours. Yes, *that* much would have to be cut before the roast went on the air (we wound up with about fifty minutes of show after commercials).

Things would occasionally get a bit raw, but by and large everybody would plow through, knowing that this would probably be their one and only chance. Greg would only redo something if it was an out-and-out disaster or if he wanted to protect someone, usually the roastee.

Center camera stayed on the speaker at the podium at all times, one side camera had a two-shot of the speaker and the roastee, and the other went searching the dais for reactions from individuals seated there. We weren't always successful in getting the camera on the guest who was laughing at a particular joke. In editing I would play the reaction tape reel without sound and grade the laughs as BIG, GOOD, MEDIUM, SMALL, or JUST A SMILE. Being video with no sound, they had nothing to do with what was being said at the time they were recorded. It was a chore to make the laughs fit the size of the jokes.

The reaction camera often caught stars off-guard — talking, coughing, picking their noses, even yawning — or as in Zsa Zsa's case, asking the person next to her what was just said. Lucy was forever taking out her little mirror and checking her hair. Dom DeLuise munched incessantly on the peanuts provided on the dais. We later took the trays of food off the dais for that reason.

Three more top stars fell in line to be roasted: Jackie Gleason, Michael Landon, and Sammy Davis Jr. They all enjoyed the sessions and there were few hitches. Sammy, particularly, felt very honored that so many friends and coperformers would show up just for him. He especially appreciated our invitation to his wife, Altovise, who introduced him. When Sammy got up for his rebuttal he was so wrapped up with the proceedings that he ignored the jokes on the cue cards and decided instead to give everyone there a warm and sincere round of gratitude. At the time we felt it was extremely out of place and a definite downer. But leaving his tearful words in the final edited show gave it a welcome, satisfying switch.

To get to Las Vegas from Los Angeles for

the roasts, Dean (without Kathy Hawn) and his friends would rent a Western Airlines jet, occasionally asking one or two of that week's guests to come along. The plane got Dean to Vegas the night before the first roast and dashed him back to L.A. immediately following Greg's wrap on the last day. His fear of elevators demanded that he take a suite on the first floor of rooms, which even then was five flights of stairs for him. Hotel guards made sure no one bothered him. Even the staff stayed away. Greg, Mort Viner, and the cue card boys were the only ones allowed in Dean's little corner.

Dean, Gene Kelly, and George Burns

The George Burns roast: (*back row*) Tom Dreeson and Charlie Callas; (*second row*) Milton Berle, Frank Welker, Ruth Buzzi, LaWanda Page, Phyllis Diller, Abe Vigoda, Jack Carter, and Dom DeLuise; (*front row*) Orson Welles, Connie Stevens, George, Dean, Gene Kelly, and Red Buttons

As isolated as he was in the hotel, Dean still revealed the strain of what was happening in his home life. Unlike his early days with us, he seemed very low, taking his sweet time to arrive for a taping, keeping stars and crew waiting, sometimes for budget-stretching minutes. It was so unlike the on-time, anything-you-say days we knew at NBC.

Kathy would call him long distance from California, somehow always timing each call to reach him just as our taping was ready to start. He would come to the stage, belatedly, in a very bad mood.

"Hang on, everybody," Greg would say from the control room, "it's going to be a rough night."

Sometimes Dean would snap out of it, sometimes not. If he didn't, it meant there'd be less of Dean in that particular show, with reactions from previous shows edited in to make him look like he was having a grand old time.

At one roast, we were all standing around waiting for him backstage when we heard a loud crashing of glass. A clumsy stagehand, we thought at first. But no, the noise came from Dean's makeshift dressing room on-stage. A call from Kathy had so infuriated him that he actually knocked over a shelf of liquor bottles.

"No close-ups tonight, boys," Greg told us.

Soon after that, separation announcements started to appear in the gossip columns. In 1976 they were finally divorced. At the next roast, I remarked to Dean how well he looked.

"Why shouldn't I?" he said. "I've just had a terrible burden taken off my back."

Muhammad Ali was an "on-and-off-er,"

Dean and Jimmy Stewart

first agreeing to be roasted, then having second thoughts. He surprised us all by actually showing up. We didn't think he would. There'd been heavy tinkering with his rebuttal, and as we all knew in those days, if Ali didn't want to do something, he didn't do it.

But there he was in Vegas, royally enjoying the preshow party backstage. He was especially kind to Dean.

"You know, you don't seem like a man who would have a big television show like this — such a world figure, nice lookin'...."

To the MGM Grand audience he tried to explain his feelings toward Dean: "Dean don't really care about these shows. He don't care nothin' about the scripts. He came to me backstage and said, 'Hi ya, champ. How ya feelin'?' I said, 'Fine.' He said, 'Got your speech down?' I said, 'Yep.' He says, 'Let's go.' I mean, this man is *serious*."

During most of the taping, though, Ali just sat there on the dais, staring out at the audience. He seemed very displeased with what everyone was saying about him. We couldn't decide whether he didn't understand that what they were saying was supposed to be funny or not. Maybe he thought it was just cruel. Or perhaps he was simply bored.

He came to life twice. During Howard Cosell's spot he jumped up and playfully tried to remove Howard's famous hairpiece.

"Why did you invite Cosell, Dean? I've made him famous enough." He obviously liked Howard.

Ruth Buzzi's "Gladys" bit went over big with Ali. Ruth originated the character when she and Dom DeLuise had an act together. Appearing often on *Laugh-In*, "Gladys" was a loser, an extremely unattractive girl with clothes to match who kept hitting people with her handbag. Ruth had worked with Ali on several other TV shows and always felt somehow that he didn't like her. She was thinking of this when she met him before the roast and informed him that he'd be hit by her purse and she hoped he wouldn't mind too much. He didn't seem to know who she was, much less understand what she was talking about, which made Ruth even more concerned about her use of the handbag on the taping. She gave him a few whacks to show that it really didn't hurt. He looked like he felt sorry for her and walked away. By this time, Ruth was really upset and decided to bombard him unmercifully when it came to her entrance. Ali enjoyed it immensely. It was the only time on the show, as a matter of fact, that he really broke up.

One of her lines really broke him up: "I begged him to give up fighting. I talked and talked until I was black in the face. But then, you wouldn't know about that."

Ruth had to be dragged offstage by Dean, Red Buttons, and Tony Orlando — not once, but three times. She kept coming back to blast Ali again.

During Rocky Graziano's speech, after Ruth had taken her seat at the very end of the dais, Ali bent over and mouthed some words to Ruth way down at the other end, something like, "Can I borrow your purse?" Rocky continued as Ali kneeled down like an elf and walked behind everybody seated at the dais to come up and grab Ruth's purse. He took it back to the podium and began slamming Rocky with it. Mission accomplished, he bent down and returned the bag to Ruth.

I used all Ali's "Gladys" reactions in the final editing, completely changing his reactions to other speakers. While he had been scowling throughout most of it, in the final edited show he laughed uproariously at everything. It was a false illusion, but I'm sure even Ali would have agreed with it. He said afterward that he'd had a good time, so maybe he thought he was smiling all the time.

Dean and
Betty White

Oh, what would we have done without editing? We made Dean look happy when he wasn't, made Muhammad Ali and others look like they were thoroughly enjoying themselves when they weren't. There were even ways of rescuing disastrous spots that had been read badly by a star. It was always cut down to the best sections, roundly cheered and applauded.

Only Orson Welles ever made mention of the little miracles we were performing in the editing room. He understood. He always thanked Greg and me for making him look good.

★ ★ ★

When Greg booked George Burns and Jimmy Stewart for roasts, he knew the network would agree to extending our playing time to ninety minutes. He also knew he wouldn't have any trouble getting guests to be on the dais. It was somewhat difficult for the writers to scorch these two lovable people. The guests felt the same way and asked that they not have to be too cruel. Their shows became more tributes than roasts. Jimmy's turned out so well that Greg asked for and received permission to make his roast a full two hours.

For Jimmy we asked as many of the old MGM contract players as we could to join us. We managed to corral Greer Garson and Henry Fonda, but only if they could tape their spots in a "piece of the roast set" we kept at NBC in Burbank, the center of the dais and the roastee's picture. Others agreed to come up to Vegas, including June Allyson, Mickey Rooney, Janet Leigh, Eddie Albert, Lucille Ball, and Orson.

My early secret love for June Allyson was almost as strong as what I had felt for Alice Faye, and Greg decided to tell that to June. In her famous wonderful scratchy voice, she said that was very flattering and suggested we have our picture taken together.

438

Mickey Rooney was standing nearby and walked over as the picture was being taken and grabbed June away from me.

"Junie's my kinda gal," he said. Then he calmly explained that when they were together as young stars at MGM, "Junie was the only girl who wouldn't let me fool around with her. And she was just my size, too. Too bad."

"Oh, you were always too busy with everybody else," June added.

"Wanna make it now, kid? It's never too late."

Jimmy Stewart, who added up to about two Mickey Rooneys, stepped in to remind Mickey that June was *his* girl in all those movies and not to forget it.

"Hell, Jimmy, you even made a movie with one of my ex-wives."

"Who didn't?" Jimmy shot back.

Orson, who seldom departed from his written words, ad-libbed something he felt sincerely about Jimmy:

"As an actor, I saw Jimmy on the stage in New York, long before Hollywood got him. No, that's wrong. Hollywood never got Jimmy Stewart. He was the conqueror. He was and is superb. And I use the word very carefully. I pause for parentheses. I'm the only actor on this podium who has not attempted to imitate you tonight, Jimmy . . . because in my view, that's all we can do."

Jimmy was very touched and didn't hesitate to say so on-camera. Naturally, I left it all in. Good stuff.

Valerie Harper had just gotten her own show, *Rhoda*, a surefire spin-off from her previous hit, *The Mary Tyler Moore Show*. Her sudden TV stardom would make a ratings high roast, especially if the entire cast of her show showed up. Most of them readily agreed (Julie Kavner, David Groh, and Harold Gould), but Nancy Walker, who played Rhoda's mother, gave us a rough time. She made a lot of demands, and negotiations were on and off like flashing neon signs. When she eventually said she'd do it, all hell broke loose.

"She wants the script changed — again!" the writers kept screaming before we even got to Las Vegas. She drove them crazy. Every time she didn't like something, she'd tell her agent to get her off the show. Greg reminded both of them that they had signed a contract and were expected to be at the Grand at the specified time, script changes or not.

Ordinarily, if a star wanted out, Greg would let them go, usually with a promise to "make it up later." But Nancy's mother role on Valerie's show was important to the roast. Viewers would wonder why she wasn't there with the others. Besides, because she had been so unreasonable, Greg decided to hold her to her signature.

She showed up, asking for still more changes, and was given all the benefits of the doubt. She sat on the dais between Valerie and Julie, looking exactly like Kim Hunter in *Planet of the Apes*, which she always did in those days. When it was her turn to speak, she decided to get back at the writers.

"I don't want to read that junk," she said as she waved the cue cards away. She ad-

Me, Orson Welles, Dean, and Greg

libbed some remarks about her TV daughter, things she seemed to feel were terribly witty, but she sat down to an ovation of silence. The audience was bewildered. You could hear them saying to each other, "What did she say?"

Greg turned to me in the control room and gave me the cutthroat gesture that meant I should take Nancy out of the edited show. Unlike others who goofed a few lines, she was not asked to remain after the audience left for "pickups."

Dr. Joyce Brothers was not only an intelligent lady, but deadly serious about whatever she was being paid for. Every time she did a roast for us, she was less concerned about *what* she was going to say than with *how* she would say it. She said she figured our writers knew better what was funny than she did. Now it was up to her to make it all work. She desperately wanted to give it her best shot. She'd ask me or Jonathan Lucas to coach her on every little inflection, every pause, every nuance. We felt like we were preparing her for a Broadway show.

She was most grateful for our help, but I always felt the end result looked like she had indeed been directed. It was our fault that it seemed phony. I hoped it was not as obvious to the viewing public as it was to me. Dr. Joyce didn't need coaching in the art of charm and good manners, though — backstage she was warm and bright, and endeared herself to cast and crew by remembering everybody's name.

Shelley Winters has been in just about every branch of showbiz, but the roasts were just far out enough to give her some problems. She always had to hang around for "pickups," usually redoing her entire spot. Sometimes a bit too much of the bubbly before the show was the main reason for this, but there was also a situation that she, like other actors, fell into. They felt they could be just as cute as the comedians on the dais and would start to ad-lib, resulting in excruciatingly unfunny remarks. Suddenly they found themselves so tangled up they couldn't get back to the script, at least not smoothly. The audience sensed the trouble and became deadly quiet.

But Shelley would hang in there. When it was explained to her that a good actress deftly reading a comic line could come off better than a tired old comedian, she'd recover her confidence and sail through.

Orson, of course, merely had to show up

and open his mouth once to gain the respect of everyone in the house. He liked doing the roasts and considered them another kind of theater to conquer, but he hated that nonsense before the show when all the stars got together to say how absolutely thrilled they were to see each other. Acknowledging that these were inevitable moments of Hollywood ego trips, he would insist that he not come down to the stage until the last possible moment. Meanwhile, Greg and I kept ourselves busy making sure everybody knew everybody else: "Gabe Kaplan, I'd like you to say hello to Hank Aaron." "George Hamilton, this is Milton Berle." "Senator Goldwater, meet Charlie Callas."

Dean, Angie Dickinson, and Earl Holliman

If Orson did happen to come down too early, he would sit quietly on the opposite side of the stage, hoping no one would notice him. He'd look so woebegone in the darkness, that imposing figure of a man, that giant of the spoken word. I think he intimidated everyone just by his appearance. Stagehands and fellow thespians left him alone. They must have been afraid he might growl, possibly bite.

One night after we finished taping I found him sitting alone again, still in his huge tuxedo, on a folding chair outside the stage door next to the Grand's big kitchens. I wondered why he hadn't gone back to his room.

"Oh, Greg wants me to go across the street to Caesar's Palace to see Sinatra. He went up to change. It's easier for me to go like this. Do you think anybody will pay any mind to that?"

"I think you're like Dean," I said. "Tuxedos look right on you. You're a classy guy."

"I just don't want to bother going all the way up to my room and back down again. At this hotel, that's a definite chore."

I grabbed another folding chair and sat down next to him to keep him company while he waited.

When Greg arrived in coat and tie, they

Me with June Allyson and Janet Leigh for the Jimmy Stewart roast

insisted I, in a short-sleeved shirt, join them. Well, Vegas is that kind of place.

We arrived in the Circus Maximus room just as the lights were dimming for comedian Pat Henry's entrance. Even in the dark, customers easily recognized Orson making his way to a front table. There were flutters of applause as we passed through.

Orson tried to make a few stabs at laughing at Henry's routine, but was much more comfortable when Sinatra came on. "This is the cream of the crop," he whispered to me. About halfway through his act,

Frank introduced Orson amid thunderous applause. He told the audience what dear old friends they were and how much he owed his career to Orson.

What he didn't tell them was that Orson coached Frank when he was preparing for *From Here to Eternity.* Frank felt that it was Orson who was responsible for his Academy Award for that role and subsequently the beginning of his important movie parts. Later, Greg asked Dean if it was true about Orson training Frank. "It's not only true," Dean quickly stated, "but did you know that Marlon Brando did the same for me on *The Young Lions?*"

After Frank's brilliant performance we were led backstage through a series of curtains and guards to Frank's dressing room, where a very attractive lady (soon to be his new wife) asked if we'd like the same thing we'd been drinking at our table. Sure enough, Barbara had made the effort to find out exactly what wine we were having in the Circus Maximus and supplied us with more of the same.

When Frank came in, Greg mentioned that we were putting together NBC's 50th anniversary show and that Orson was going to narrate it.

"They never let me see anything," Orson

grumbled. We had recorded Orson's voice-overs for the NBC show, and Greg merely told Orson what the sections were about without showing them to him.

I told Frank that we were going to use a lot of clips from his early television performances and he was very interested in what we had picked. I sensed considerable concern.

"Do you remember your very first TV performance?" I asked him.

"Do I? I was a nervous wreck. It was something called *Star Spangled Revue*, I think. What did I sing?"

" 'Come Rain or Come Shine.' "

"Right. How was I?"

"Terrific."

Greg told him we were going to include the long medley he did with Dean on our Christmas show.

"Oh boy, was he great then." Frank smiled, then switched to a little frown.

"I hope you have that clip of Martha Raye carrying me off stage on *The Milton Berle Show*. Of course, I was thinner then."

He poured us another round of wine.

"You'll use the fun stuff, won't you?"

I assured him we would.

"You guys did a great job with Dean. I know you'll be just as kind to me. Would you like to stay for the second show?"

Some people have style. Some have class. Sinatra had both.

In the middle of the 1975-1976 season, we decided to do something we'd been wanting to do for a long time — turn the tables and roast Dean.

"It's got to be a super show," Greg insisted, "the most stars, the best writing — we'll talk NBC into giving us two hours." Dean was in a great mood after the divorce from Kathy and we figured we'd best hurry and capture that.

The cast certainly was super: Bob Hope, Paul Lynde, Angie Dickinson, Rich Little, Tony Orlando, Joe Namath, Joey Bishop, Senator Hubert Humphrey, Senator Barry Goldwater, Foster Brooks, Ruth Buzzi, Jimmy Stewart, Gene Kelly, Orson Welles, Charlie Callas, Nipsey Russell, Rowan and Martin, Gabe Kaplan, Howard Cosell, Georgia Engel, John Wayne, and Muhammad Ali. Because so many stars were involved and their availabilities were so complex, it seemed easier to tape in L.A. rather than fly them all to Las Vegas. Consequently, the entire MGM Grand set was trucked in and set up at the ABC studios in Hollywood.

Dressing rooms for so many people became our major headache. On that kind of roster everybody deserved a "star's" room but there weren't more than a couple of those at ABC. We grabbed everything they had, big and small, plus a couple of trailers, and it still wasn't enough. We asked the celebrities to double- and triple-up and they all willingly obliged, except Redd Foxx, who arrived with an ultimatum: either he was to get the best dressing room or he was walking. Associate producer Roger Warnix tried to explain our problems, telling Redd that everyone else was cooperating, but he held firm.

"Even Humphrey and Goldwater are sharing a room," Roger pleaded.

"I don't give a damn about those honkies," Redd shot back. By the time all this got to Greg, Redd was headed for the exits.

"I'm sorry, Greg," Roger said. "I did everything I could but he was real nasty about it."

"It's okay, kid. But try one more thing. Run out to Redd's car and tell him that he can have the Green Room. We'll kick out Hope and Kelly. See what he says."

Roger returned to say Redd had disappeared.

"To hell with him!" Greg screamed.

"There's plenty of show without him." And there was. So much so that some choice gems from Don Rickles, who hosted the affair, had to be edited out to make room for everybody else. Rickles was at his best, giving the needle to everyone, no holds barred:

"Tonight we're honoring our man of the hour . . . uh, Dean Martin. You heard the applause, Dean. I'd worry."

"Jimmy! Over here, Mr. Stewart. I spoke to the family. You're doing well. (LOUDER) I spoke to your wife, Jimmy. She's leaving you."

"Orson is Moby Dick. Water comes out of his navel."

"Come on, Senator Goldwater . . . I

Dean being roasted by Bob Hope

laughed when *you* lost."

"Good to see Angie Dickinson sitting next to Joe Namath. When I blow the whistle, HIKE!"

"Angie's show, *Police Woman*, No. 4 in the country . . . so you see how our country's going."

"Hubert Humphrey just said, 'There's something wrong with the country?' "

"Goldwater said, 'Forget it, I've had my shot.' "

And while Muhammad Ali was at the podium: "Who's gonna tell him he's been on too long?"

Ali asked Don if he was telling him to get off. "No, no. This man said it. (Pointing to Orson). I'm with you, baby. I'm gonna live in Memphis with you. I'll drive the bus and *you* go to school."

Ali couldn't resist saying, "You're not as dumb as you look, boy."

Betty White had been so successful on *The Mary Tyler Moore Show* that she got her own spin-off and everybody was sure it was going to be a big hit. And we relished the thought of how wonderfully acid Betty could be in her rebuttal if we roasted her.

We taped her roast before her own show even got on the air, and unfortunately

Dean and Greg

Betty's sitcom skied down the Nielsen slopes in short order. It was cancelled before our roast of her got scheduled on NBC, leaving us with a roast about somebody whose new show had just been dropped.

That in itself would have been something we could have worked with, if we had known it was going to happen, but instead our show had nothing but rave things to say about "Betty's big new hit."

NBC kept delaying her roast, hoping her show would get a reprieve. But no, CBS threw it off and we were stuck. I had to go back in the editing room and take out as many "What a hit!" references as I could. Her roast finally made the NBC schedule seven months later.

No such problem with roastee Frank Sinatra. I guess you could never cancel him.

Once again, NBC let us make it a two-hour roast.

Among another stellar list of top names was Peter Falk, whose fifteen minute spot was so genuinely funny we left it all in, at the expense of some of the other names.

Peter came out of the audience in his Lt. Columbo character, saying he wanted to honor "another Italian." He even brought some homemade lasagna. Much of the success of the spot was due to the clever writing (about how he knew Sinatra in Brooklyn and was so pleased that Frank would remember him), but a lot of it was Peter's delivery. It broke up everybody, especially Frank.

Incredibly, after it was over we couldn't convince Peter it had been so great. He talked Greg into redoing the whole spot, without the audience and without Dean and Frank.

"I can make it better, I know I can." We painstakingly went through it all once more with many additional pickups until Peter was reasonably happy. We taped it with Greg sitting in for Frank and me for Dean, so he'd at least have bodies to react to. A perfectionist, he wanted the end result to be his best.

"I don't want to walk away short," he kept saying. But you can guess what happened.

We totally ignored the endless retakes and used the original tape.

To TV viewers watching the roasts, Dean Martin was the same hip, wonderful guy they'd known since the early '50s. But we saw changes, mean ones. His timing was off, this from a man whom even Jerry Lewis had called the world's greatest straight man. He would often wander aimlessly away from the cue cards, sometimes belittling the audience for not laughing hard enough, insulting them so badly they must have figured it was part of his act, that he couldn't possibly mean what he was saying. Still, they weren't sure — and it made it that much harder for the next person up on the dais.

We were having more and more trouble getting good reactions from him on-camera. We finally went back a couple of years and found some old shots of him and used those laughs and smiles.

Worst of all, there was that famous red handkerchief, which he began to take out of his pocket and sniff from time to time. What was in that hankie? We could only guess, but it couldn't be good.

The old happy-go-lucky, camera-wise, jolly Dean Martin was disappearing right in front of our eyes.

13 The Specials:

Not Like the Old Days

In 1975 critics suggested it was time for Dean to get back to a weekly series, but that wasn't how NBC saw it. They wanted more roasts. As for Dean doing another variety series, the network biggies were quick to snap, "He's too old-fashioned."

There'd been a big changeover at the network. The average age at the top was down to around thirty, and those were the decision-makers.

"We're grooming Mac Davis to take his place," the young new executives told Greg.

To us, Mac Davis was a very talented performer, but he was no Dean Martin. Dean old-fashioned? We always thought he was pretty hip. But just in case, we'd tried to surround him with new faces, a sort of "something for everyone" attitude. One thing was for sure, we didn't want to do another "old-fashioned" Christmas special in 1975. We decided to go outdoors. "Christmas in Cali-

456

fornia," I argued, was different from Christmas anywhere else. "People barbecue their turkey outside. Instead of a sleigh ride, they ride horseback. Rather than throwing snowballs, they toss Frisbees at the beach."

As Irving Berlin put it in his verse to "White Christmas":

The sun is shining, the grass is green,
The orange and palm trees sway.
There's never been such a day
In Beverly Hills, L.A.

We took it from there. Taping was scheduled for the week before Thanksgiving and the weatherman came through. It was beautifully warm and sunny.

"Nobody will believe we didn't do this show last summer," Greg smiled.

Our first location was Leo Carillo Beach, a state park just 50 miles northwest of Burbank, full of picturesque rocks, ocean views and sand dunes. Crew and cast (except Dean) arrived at 5 a.m. to set up.

Dionne Warwick was one of our guests but didn't have time on her schedule to rehearse in Burbank, so this was our first meeting. I had communicated with her earlier by long-distance, explaining what she'd be doing. I knew Dionne was one of the few

Dionne Warwick, Michael Learned, Dean, and Georgia Engel at Greg's ranch for a Christmas special

stars around who didn't need a lot of rehearsal. She was a marvelous musician by instinct, took direction quickly, and was as professional as they come.

In the early morning darkness on the beach, dressers and makeup people were getting our guest stars prepared. The cast included Michael Learned, Georgia Engel, Freddie Fender, the Statler Brothers, and the Golddiggers. The camera crew was out on the beach with flashlights getting set for the first shots. I sat with Dionne as she was getting made up, going through her parts with a script and an audio cassette. Besides her solo, she had a medley with Dean, two numbers with the other ladies, a production number on the beach, and a finale at Greg's ranch in Hidden Valley. No problem with Dionne. In a very short time she was as ready as the rest of the cast, who'd been rehearsing for over a week.

Just as the sun came up and we were set up for our first shot, there was a disturbing call from Mort Viner. Dean had "a touch of the flu" and wouldn't be able to make it to the location.

This was definitely not like Dean. We fully expected him to show up on time or earlier and be ready to tape. We had put him in just about every number, so it would be difficult to work around him. Nevertheless, with everybody waiting for the words "roll tape," we felt we had to put something in the can.

It was the only day scheduled for the beach; the rest of the show would be done at the ranch.

A touch of the flu? How odd. Dean used to show up no matter what. The staff got together for a brief meeting to see what we could salvage without him. Once again I stood in, the cameras making sure I would not be seen as the rest of the cast went through the paces. Somehow we'd have to get Dean back there on the beach for close-ups and put it all together. It wouldn't be easy.

The next day at the ranch he showed up, seeming to have made a miraculous recovery, and was hot to go. We raced through that day's schedule and in the afternoon whisked him off to the beach to fill in the blanks from the day before.

Michael Learned was reluctant to do a musical show, saying it was totally out of her field. But after a few sessions at the piano I discovered she had a lovely singing voice and I was able to talk her into a solo on "You've Got a Friend," as well as numbers with Dean, Dionne, and the rest of the cast.

"Just don't have me singing next to Dionne," she said. "My inadequacies will be in full view of the entire American public."

Dean's Place special — Dean and the Golddiggers

The 1980
Golddiggers

She rose to the occasion brilliantly and showed a particularly nice romantic tendency with Dean.

"Don't let it fool you," she said later. "It was only the thrill of the moment — all that music and natural scenery and — well, he's a charmer, isn't he?"

Although it was pieced together with hope and prayers because of Dean's no-show that one day, the show got unusually good reviews:

Dean turned all the traditional Christmas shows topsy-turvy in his own inimitable style and by golly, it worked.

Martin dreams of a 'White Christmas' while the rest of the country, fed up with the cold, is probably on their way West.

461

We did several other specials that year, one of which was a feeble attempt to get Dean back into the variety look. We took over a nightclub in West Hollywood and re-named it *Dean's Place*. His special guests included Robert Mitchum, Angie Dickinson, Foster Brooks, Sherman Hemsley, Isabel Sanford, Jessi Colter, Jack Cassidy, Milton Berle, the Untouchables, the Golddiggers, and Nancy and Ronald Reagan.

We weren't confronted by any flu symptoms this time, but there were other obstacles. Dean had a fancy trailer for a dressing room. He was still married to Kathy Hawn at the time and she shared his trailer during all the rehearsals, carefully watching everything we did. Their arguments were easily

Dean and the Golddiggers at Greg's ranch for a Christmas special

Dean and Jonathan Winters at Greg's ranch

heard by cast and crew and somehow re-peated to the press. The gossip columnists had a field day.

We tried to ignore it all, except one big storm had to do with some choice com-ments Kathy made about the Golddiggers and how they were getting a little too cozy with Dean in their numbers. And perhaps Dean was enjoying it too much. Everybody heard that one, and we knew there'd be some sort of ultimatum from the trailer.

Greg decided to let the taping go as it was and work it out in the editing room, which resulted in a whole number being removed from the show. "A personal request," I was told. Kathy had made her objections so strong that to keep harmony at home, Dean had asked to eliminate his medley with the

Golddiggers. It was unlike Dean ever to interfere. We could only assume that he was pressured into it by his young wife.

The girls were shocked. They were only doing what the Golddiggers had always done with Dean, having a little innocent fun. They thought the whole mess was silly, and so did the rest of us.

Two tries with *Dean's Place* didn't result in any variety resurgence, but we did make one more stab at it with something called *The Red Hot Scandals of 1926*. Instead of the '30s, where we had gone for the early Golddiggers shows, this time we went back still further for a look at the '20s with Hermione Baddley, Dom DeLuise, Abe Vigoda, Charlene Ryan, Georgia Engel, and the Golddiggers. We took over some sets at Warner Brothers for collegiate hijinks, a Chicago speakeasy, and a look at old Hollywood with the change from silent films to talkies. Two segments used the Biltmore hotel in downtown Los Angeles for a typical '20s radio show. Saving us all was Jonathan Winters, who brilliantly characterized Al Capone, Charles Lindbergh, Knute Rockne, and Douglas Fairbanks.

The finales were all shot at the Music Hall in Santa Monica, which we didn't have to change much because it already looked like the '20s. Having rid himself of Kathy by that time, Dean was in good spirits and worked well with the rest of the cast, especially Jonathan. Hermione gave us a few anxious moments, however, seldom showing up after lunch (which obviously included several vodka stingers) and causing a reshuffling in the schedule.

In spite of the fancy trappings, it was two *Red Hot Scandals* and out. "Stop the music and give us more roasts!" NBC demanded.

In May 1977, a 4,000-seat theater in Westchester County, New York, was sold out for three weeks after one small ad ran in the Sunday *New York Times*. The occasion: concerts by Dean Martin and Frank Sinatra.

Everybody expected the shows would do well, but complete sell-outs after one tiny announcement? We knew the old-timers would be there, the fans of both stars from their early days, but the kids were there, too!

They came early, hoping to get a glimpse of the two showbiz giants. Security was the most elaborate that that part of the country had ever seen.

Opening night was unbelievable, totally unexpected. When the announcer said, "And now . . . direct from the bar . . ." the

Dean, Lynn Anderson, and Buck Owens at San Diego's Sea World for the 1981 Christmas special

shouting began, ". . . Dean Martin!" The din began to swell. It got louder and louder. Some adoring fans started toward the stage. The guards held them back.

Dean shook his head, unable to grasp the onslaught of adoration.

It took quite a while before they calmed down enough for him to utter his standard opening line: "How long have I been on?"

The audience cheered for that and every line after it. They went bananas over every one of his songs. He stayed on for half an hour, then introduced Frank, who did thirty-five minutes with more song and less patter than Dean.

As Frank finished his last song, Dean

Dean, Beverly Sills, Mel Tillis, Andy Gibb, and Eric Estrada at NBC for the 1982 Christmas special

rolled his portable bar on stage and the two of them started toasting everyone they could think of — their friends in show business, the band, the audience, each other. This is what the audience had really come for, to see these two superstars together again — ad-libbing, drinking, joking, singing, doing what *they* did so well.

Forty minutes later, as the band struck up "I Write the Songs," they strolled off the stage. But the audience wasn't about to let them go. They stormed to the front, reached for the two startled Italian singers, tried to touch them.

Dean and Frank couldn't believe what was happening. They instinctively went to the apron of the stage and shook the out-stretched hands, supported more often than not by some muscular security guards.

The same scene was repeated at the exact same time in every performance after that. It wasn't planned or staged. It just happened.

"DEAN MARTIN AND FRANK SINATRA — ROCK STARS?" headlined the *New York Times*.

Yes, there was still some magic there. The mystique, the wonder of those two super-relaxed singers of America's songs doing what they do best and pleasing everybody, including themselves, was still alive.

In the limo driving back to the hotel, Dean was in tears. "I can't believe it! I can't believe it! They remembered me."

For one brief moment, we thought maybe NBC would take note of this phenomenon and let us get out the band and the dancing shoes. But no, the baby boomers in charge insisted on more roasts.

They did allow us an annual holiday show, however, with two more excursions to Greg's Hidden Valley ranch, two at San Diego's Sea World, one at the Wild Animal

Park, and another at the San Diego Zoo, continuing our *Christmas in California* theme. The results were diminishing. Each year was a struggle to make Dean look good. He was always right there and ready, accepting anything we suggested. The spirit was willing, but Dean's age was catching up on him. On one of the shows at Sea World, we weren't even sure Dean would be able to carry off a song. The voice was wobbly and unsure.

We dug out some of his records and asked him to lip-sync them. He didn't argue about it, apparently agreeing that some of the magic was gone. We all felt for him.

Dean's last Christmas show, this time deliberately safe back in the NBC-Burbank studios in 1981, was nearly a disaster. The voice was still not good, but at least we were where we could fuss with the audio and do second or third takes if we had to. Mel Tillis had worked with Dean several times before and knew how to handle him, but his other guests — Beverly Sills, Andy Gibb, and Eric Estrada — remembered Dean from his early days and were aghast to find him incoherent and vague.

"He's killing himself," they said to each other. Beverly guided him like a mother through all they did together, trying her best not to show her concern over his health.

Andy was more excited than anyone about working with Dean, his idol since he grew up in Australia. He couldn't wait to meet him. Although Dean was pretty much out of it, Andy still showed his reverence. He and Beverly and the rest of the cast eventually wound up treating Dean gently, like an old friend who was not taking very good care of himself. They protected him and helped him along. But they were shocked.

A year later Dean seemed to be coming around a bit, but it was too late. There were three more roasts — Joan Collins, Mr. T., and another round with Michael Landon — but no more musical outings. We more or less knew that the network was testing Dean with those three shows. They were not impressed with the ratings, even with big-name guests. They officially announced that there would be no more Dean Martin shows, not even roasts. "Nobody's interested anymore," they told Greg.

It was over.

14 Finale

I wonder where all of us who worked on *The Dean Martin Show* would be today if Dean's health hadn't slowed him down.

"We'd still be on the air," is Greg's optimistic answer.

Although that seems highly unlikely, it's true that week after week during all those years, we watched a unique entertainer. Dean could do it all. He was a sensational singer of popular songs. His comedy timing was better than anyone's in the business. He was kind, compassionate, and considerate of all around him. He was a man's man and a lady's dream.

And who else has been so successful in so many fields? He was a hit in nightclubs, movies, on records, and on television. And he knew the remarkable secret of making it all work — he never took himself too seriously.

No wonder *The Dean Martin Show* did so well. We had the perfect television performer.

Orson Welles called him "the world's second-greatest maverick," insisting he himself was the first. Kate Smith treated him like a naughty son, verbally spanking him for his smoking and drinking, but loving him very much. Peggy Lee and Lena Horne insisted there was no easier man to sing with. Dom DeLuise, Charles Nelson Reilly, and even Paul Lynde, like Jerry Lewis, considered Dean their best straight man. Jimmy Stewart, a casual guy in his own right, was awed by Dean's ability to drop in the studio, unaware of what was ahead for him, and make the show so much fun. Irving Berlin, one of Dean's biggest fans, told everyone that Dean sang "White Christmas" and all of his songs better than anyone else.

"Variety" is a dirty word on television today, except for a few moments on the talk shows. And music videos, popular as they are, are just that: highly produced, end-to-end music pictures with no heart.

No, today's viewers can't get what we had in the '60s and '70s — singers, dancers, clowns, actors — all rolled up in something called variety shows.

In the early '80s, twenty-six of Dean's shows went into syndication and found a whole new audience. Kids found themselves associating with this fellow who broke all the rules. "Hey man, this guy is cool. He breaks me up."

What a shame there's no room anymore

for the singers, the dancers, the clowns. It's too bad that we can't still see the wonderful *spirit* of those variety shows. Their only purpose was to make us smile, to entertain us with song and dance. Were they just a product of a different time?

Well, we still have our memories. The vision of that good-looking, classy guy in a well-fitted tux, bouncing around the stage with the world's best-known performers, joking about his favorite subjects, booze and broads, every Thursday night at ten, still makes some of us smile.

We won't forget. I miss him very much. Thanks, Dean.

Epilogue

After *The Dean Martin Show* came to an end, Dean continued to entertain America, making appearances in movies and on television. On Christmas Day, 1995, he died of acute respiratory failure at his Beverly Hills home. Ex-wife Jeanne was by his side. During Dean's last years, the two of them spent much time together, always supporting each other and their family, which now consists of thirty-seven children, grandchildren, and spouses.

Jeanne still lives in Beverly Hills and is active with nonprofit organizations such as SHARE, Inc., which she founded. She is now the matriarch of the family, which remains, as it always has been, extremely close. Jeanne's house is home for the holidays, and Gail Martin does the Thanksgiving festivities. It's a very special and loving clan.

As for Dean's children, while many of them initially pursued singing and acting careers, none has continued in their father's footsteps. Craig has now retired to Palm Springs with his wife, Donna. Claudia lives in Reno with her husband, Jim Roberts. Gail is married to *Los Angeles Times* columnist Mike Downey, and they live in Sherman Oaks, California, with their three lovely daughters. Deana is a very successful personal trainer living in Los Angeles with husband Jim Griffith. Dean-Paul, a captain in the U.S. Air National Guard, died tragically during a training mission on March 20, 1987. He is survived by Alexander, his son by actress Olivia Hussey. Ricci currently resides in Park City, Utah, with his wife, Annie, and three beautiful young daughters. Gina lives above Boulder, Colorado. Sadly, her husband, Carl Wilson of Beach Boys fame, died of cancer in 1998. Sasha, Dean's adopted daughter from his marriage to Kathy Hawn, is very happily married and living in Las Vegas. Jeanne considers Sasha a part of the family.

Those of us who put our hearts and souls into making Dean's show such a remarkable success continued in our show biz careers, working on various projects. I served as producer or musical director on many TV specials, including several starring Bob Hope, Gene Kelly, Bob Newhart, Jonathan Winters, Rowan & Martin, and Dom DeLuise, and I wrote special material for many other programs, including two Emmy Awards shows. I'm now semi-retired in Beverly Hills, producing shows for the Society of Singers and the Professional Dancers Society.

Greg Garrison is also semi-retired, living on his ranch in Montana happy to be selling *The Dean Martin Celebrity Roasts*. Associate producer Janet Tighe, a full retiree, lives in Northridge, California, where she rode out the 1994 earthquake. Pianist Geoff Clarkson claims to be "sort of retired" in Toluca Lake, California, where he lives with wife Yoriko. He spends his time writing songs and conducting the orchestra for Bob and Dolores Hope.

Others have continued in the business to this day. Les Brown's Band of Renown still plays gigs, though Les rarely conducts these days, passing the baton to his talented son, Les Brown Jr., who tours the country with the band. Les and his wife live in Pacific Palisades, California. Van Alexander remains one of Hollywood's top music arrangers. He lives with Beth, his wife of sixty-one years, in Westwood, California. Singer Melissa Stafford is very active in television, clubs, commercials, and anywhere else someone needs a lovely blonde vocalist. She married jazz pianist / composer Dave Mackay and lives in Van Nuys, California. Kevin Carlisle, the show's original choreographer, still works on staging major musical acts and television shows in Hollywood. His replacement, Bob Sidney, is active in California dance circles and serves as chairman of the advisory board of the Professional Dancers Society. Dean's longtime manager, Mort Viner, works as an executive at International Creative Management.

In addition to Dean, we have lost a few other old colleagues from the show. The man who overbooked the show in its early days, Hal Kemp, died in 1969 of liver failure. Ed Kerrigan, who started out choreographing *The Golddiggers* before moving on to Dean's show, died of cancer in 1994. Accompanist Ken Lane, on whose piano Dean jumped every week, passed away in 1996 at his home in the Lake Tahoe area. He is survived by his wife, Georgie.

Finally, the Golddiggers, some seventy-five of them throughout the years, are everything from housewives to career women. Some are in television production, some have successful dance studios, some have talented offspring in show business, and quite a few are still performing. They are all just as gorgeous as ever.

The employees of Thorndike Press hope you have enjoyed this Large Print book. All our Large Print titles are designed for easy reading, and all our books are made to last. Other Thorndike Press Large Print books are available at your library, through selected bookstores, or directly from us.

For information about titles, please call:

(800) 223-1244
(800) 223-6121

To share your comments, please write:

Publisher
Thorndike Press
P.O. Box 159
Thorndike, Maine 04986